STUDY GUIDE

Wuthering Heights
Emily Brontë

WITH CONNECTIONS

HOLT, RINEHART AND WINSTON
Harcourt Brace & Company
Austin · New York · Orlando · Atlanta · San Francisco · Boston · Dallas · Toronto · London

Staff Credits

Associate Director: Mescal Evler

Manager of Editorial Operations: Robert R. Hoyt

Managing Editor: Bill Wahlgren

Executive Editor: Emily Shenk

Editor: M. Kathleen Judge

Editorial Staff: *Assistant Managing Editor,* Marie H. Price; *Copyediting Manager,* Michael Neibergall; *Senior Copyeditor,* Mary Malone; *Copyeditors,* Joel Bourgeois, Gabrielle Field, Suzi A. Hunn, Jane M. Kominek, Millicent Ondras, Theresa Reding, Désirée Reid, Kathleen Scheiner; *Editorial Coordinators,* Jill O'Neal, Mark Holland, Marcus Johnson, Tracy DeMont; *Assistant Editorial Coordinator,* Summer Del Monte, Janet Riley; *Support Staff,* Lori De La Garza; *Word Processors,* Ruth Hooker, Margaret Sanchez, Kelly Keeley, Elizabeth Butler, Gail Coupland

Permissions: Lee Noble, Catherine Paré

Design: *Art Director, Book & Media Design,* Joe Melomo

Image Services: *Art Buyer Supervisor,* Elaine Tate

Prepress Production: Beth Prevelige, Joan Lindsay

Manufacturing Coordinator: Michael Roche

Electronic Publishing (the Connections): *Senior Manager,* Carol Martin; *Administrative Coordinators,* Rina Ouellette, Sally Williams; *Operators,* JoAnn Brown, Lana Kaupp, Christopher Lucas

Copyright © by Holt, Rinehart and Winston

All rights reserved. No part of this publication may be reproduced or transmitted in any form or by any means, electronic or mechanical, including photocopy, recording, or any information storage and retrieval system, without permission in writing from the publisher.

Teachers using HRW LIBRARY may photocopy blackline masters in complete pages in sufficient quantities for classroom use only and not for resale.

Cover: Joe Cornish/Tony Stone Images, Lois and Bob Schlowsky/Tony Stone Images

HRW is a registered trademark licensed to Holt, Rinehart and Winston.

Printed in the United States of America

ISBN 0-03-095774-5

07 08 09 10 018 08 07 06 05

TABLE *of* CONTENTS

FOR THE TEACHER

Using This Study Guide 3
Reaching All Students 4
Assessment Options 5
About the Writer 6
About the Novel: Literary Context / Critical Responses 8
The Novel at a Glance 11
Introducing the Novel 13
Plot Synopsis and Literary Elements 14

Resources Overview .... inside front cover	Answer Key 87

FOR THE STUDENT

Reader's Log 34
Double-Entry Journal 36
Group Discussion Log 37
Glossary and Vocabulary 38
Chapters I–III: Making Meanings / Reading Strategies / Novel Notes / Choices 42
Chapters IV–IX: Making Meanings / Reading Strategies / Novel Notes / Choices 46
Chapters X–XVII: Making Meanings / Reading Strategies / Novel Notes / Choices 50
Chapters XVIII–XXV: Making Meanings / Reading Strategies / Novel Notes / Choices 54
Chapters XXVI–XXXIV: Making Meanings / Reading Strategies / Novel Notes / Choices 58
Novel Review 62
Literary Elements Worksheets: Symbolism / Multiple Narrators / Indirect Characterization /
Foreshadowing / Gothic Literary Elements 64
Vocabulary Worksheets 69
Novel Projects: Writing About the Novel / Cross-Curricular Connections /
Multimedia and Internet Connections 73
Introducing the Connections 76
Exploring the Connections: Making Meanings
    Early Autumn, by Langston Hughes 78
    If the Stars Should Fall, by Samuel Allen (Paul Vesey) *and* Sorrow Is the Only Faithful One, by Owen Dodson 78
    I see around me tombstones grey, by Emily Brontë 79
    "Mr. Bell's" *Wuthering Heights,* from the *Britannia* and the *Examiner* 79
    "Reader, I Married Him," by Daniel Pool 80
    Heston Grange, from *All Creatures Great and Small,* by James Herriot 80
    The Unquiet Grave, anonymous 80
    The Bridal Pair, by Robert W. Chambers 81
    The Question, by Pablo Neruda 81
Test 82

Using This Study Guide

Approaching the Novel
The successful study of a novel often depends on students' enthusiasm, curiosity, and openness. The ideas in **Introducing the Novel** will help you create such a climate for your class. Background information in **About the Writer** and **About the Novel** can also be used to pique students' interest.

Reading and Responding to the Novel
Making Meanings questions are designed for both individual response and group or class discussion. They range from personal response to high-level critical thinking.

Reading Strategies worksheets contain graphic organizers. They help students explore techniques that enhance both comprehension and literary analysis. Many worksheets are appropriate for more than one set of chapters.

Novel Notes provide high-interest information relating to historical, cultural, literary, and other elements of the novel. The **Investigate** questions and **Reader's Log** ideas guide students to further research and consideration.

Choices suggest a wide variety of activities for exploring different aspects of the novel, either individually or collaboratively. The results may be included in a portfolio or used as springboards for larger projects.

Glossary and Vocabulary (1) clarifies allusions and other references and (2) provides definitions students may refer to as they read. The **Vocabulary Worksheets** activities are based on the Vocabulary Words.

Reader's Log, Double-Entry Journal, and **Group Discussion Log** model formats and spark ideas for responding to the novel. These pages are designed to be a resource for independent reading as well.

Responding to the Novel as a Whole
The following features provide options for culminating activities that can be used in whole-class, small-group, or independent-study situations.

Novel Review provides a format for summarizing and integrating the major literary elements.

Novel Projects suggest multiple options for culminating activities. Writing About the Novel, Cross-Curricular Connections, and Multimedia and Internet Connections propose project options that extend the text into other genres, content areas, and environments.

Responding to the Connections
Making Meanings questions in Exploring the Connections facilitate discussion of the additional readings in the HRW LIBRARY edition of this novel.

This Study Guide is intended to
- *provide maximum versatility and flexibility*
- *serve as a ready resource for background information on both the author and the book*
- *act as a catalyst for discussion, analysis, interpretation, activities, and further research*
- *provide reproducible masters that can be used for either individual or collaborative work, including discussions and projects*
- *provide multiple options for evaluating students' progress through the novel and the Connections*

Literary Elements
- plot structure
- major themes
- characterization
- setting
- point of view
- symbolism, irony, and other elements appropriate to the title

Making Meanings Reproducible Masters
- First Thoughts
- Shaping Interpretations
- Connecting with the Text
- Extending the Text
- Challenging the Text

A **Reading Check** focuses on review and comprehension.

The Worksheets Reproducible Masters
- Reading Strategies Worksheets
- Literary Elements Worksheets
- Vocabulary Worksheets

Reaching All Students

Because the questions and activities in this Study Guide are in the form of reproducible masters, labels indicating the targeted types of learners have been omitted.

Most classrooms include students from a variety of backgrounds and with a range of learning styles. The questions and activities in this Study Guide have been developed to meet diverse student interests, abilities, and learning styles. Of course, students are full of surprises, and a question or activity that is challenging to an advanced student can also be handled successfully by students who are less proficient readers. The interest level, flexibility, and variety of these questions and activities make them appropriate for a range of students.

Struggling Readers and Students with Limited English Proficiency: The **Making Meanings** questions, the **Choices** activities, and the **Reading Strategies** worksheets all provide opportunities for students to check their understanding of the text and to review their reading. The **Novel Projects** ideas are designed for a range of student abilities and learning styles. Both questions and activities motivate and encourage students to make connections to their own interests and experiences. The **Vocabulary Worksheets** can be used to facilitate language acquisition. **Dialogue Journals,** with you the teacher or with more advanced students as respondents, can be especially helpful to these students.

Advanced Students: The writing opportunity suggested with the **Making Meanings** questions and the additional research suggestions in **Novel Notes** should offer a challenge to these students. The **Choices** and **Novel Projects** activities can be taken to advanced levels. **Dialogue Journals** allow advanced students to act as mentors or to engage each other intellectually.

Auditory Learners: A range of suggestions in this Study Guide targets students who respond particularly well to auditory stimuli: making and listening to audiotapes and engaging in class discussion, role-playing, debate, oral reading, and oral presentation. See **Making Meanings** questions, **Choices,** and **Novel Projects** options (especially **Cross-Curricular Connections** and **Multimedia and Internet Connections**).

Visual/Spatial Learners: Students are guided to create visual representations of text scenes and concepts and to analyze films or videos in **Choices** and in **Novel Projects.** The **Reading Strategies** and **Literary Elements Worksheets** utilize graphic organizers as a way to both assimilate and express information.

Tactile/Kinesthetic Learners: The numerous interactive, hands-on, and problem-solving projects are designed to encourage the involvement of students motivated by action and movement. The projects also provide an opportunity for **interpersonal learners** to connect with others through novel-related tasks. The **Group Discussion Log** will help students track the significant points of their interactions.

Verbal Learners: For students who naturally connect to the written and spoken word, the **Reader's Logs** and **Dialogue Journals** will have particular appeal. This Study Guide offers numerous writing opportunities: See **Making Meanings, Choices, Novel Notes,** and **Writing About the Novel** in **Novel Projects.** These options should also be attractive to **intrapersonal learners.**

Assessment Options

Perhaps the most important goal of assessment is to provide feedback on the effectiveness of instructional strategies. As you monitor the degree to which your students understand and engage with the novel, you will naturally adjust the frequency and ratio of class to small-group and verbal to nonverbal activities, as well as the extent to which direct teaching of reading strategies, literary elements, or vocabulary is appropriate to your students' needs.

If you are in an environment where **portfolios** contain only carefully chosen samples of students' writing, you may want to introduce a second, "working," portfolio and negotiate grades with students after examining all or selected items from this portfolio.

The features in this Study Guide are designed to facilitate a variety of assessment techniques.

Reader's Logs and Double-Entry Journals can be briefly reviewed and responded to (students may wish to indicate entries they would prefer to keep private). The logs and journals are an excellent measure of students' engagement with and understanding of the novel.

Group Discussion Log entries provide students with an opportunity for self-evaluation of their participation in both book discussions and project planning.

Making Meanings questions allow you to observe and evaluate a range of student responses. Those who have difficulty with literal and interpretive questions may respond more completely to **Connecting** and **Extending**. The **Writing Opportunity** provides you with the option of ongoing assessment: You can provide feedback to students' brief written responses to these prompts as they progress through the novel.

Reading Strategies Worksheets, Novel Review, and Literary Elements Worksheets lend themselves well to both quick assessment and students' self-evaluation. They can be completed collaboratively and the results shared with the class, or students can compare their individual responses in a small-group environment.

Choices activities and writing prompts offer all students the chance to successfully complete an activity, either individually or collaboratively, and share the results with the class. These items are ideal for peer evaluation and can help prepare students for presenting and evaluating larger projects at the completion of the novel unit.

Vocabulary Worksheets can be used as diagnostic tools or as part of a concluding test.

Novel Projects evaluations might be based on the degree of understanding of the novel demonstrated by the project. Students' presentations of their projects should be taken into account, and both self-evaluation and peer evaluation can enter into the overall assessment.

The **Test** is a traditional assessment tool in three parts: objective items, short-answer questions, and essay questions.

Questions for Self-evaluation and Goal Setting

- What are the three most important things I learned in my work with this novel?
- How will I follow up so that I remember them?
- What was the most difficult part of working with this novel?
- How did I deal with the difficulty, and what would I do differently?
- What two goals will I work toward in my reading, writing, group, and other work?
- What steps will I take to achieve those goals?

Items for a "Working" Portfolio

- reading records
- drafts of written work and project plans
- audio- and videotapes of presentations
- notes on discussions
- reminders of cooperative projects, such as planning and discussion notes
- artwork
- objects and mementos connected with themes and topics in the novel
- other evidence of engagement with the book

For help with establishing and maintaining portfolio assessment, examine the **Portfolio Management System** *in* ELEMENTS OF LITERATURE.

Answer Key

The Answer Key at the back of this guide is not intended to be definitive or to set up a right-wrong dichotomy. In questions that involve interpretation, however, students' responses should be defended by citations from the text.

About the Writer
Wuthering Heights

More on Brontë

Benvenuto, Richard. **Emily Brontë.** Boston: Twayne Publishers, 1982.

Fraser, Rebecca. **Brontës: Charlotte Brontë and Her Family.** New York: Fawcett Books, 1990. A fresh new look at Charlotte Brontë and the way she helped to change societal perceptions about women.

Kavanaugh, James H. **Emily Brontë.** New York: Basil Blackwell, 1985.

Spark, Muriel, and Derek Stanford. **Emily Brontë: Her Life and Work.** London: P. Owen, 1960.

Also by Brontë

The Complete Poems of Emily Brontë, ed. by C. W. Hatfield. New York: Columbia University Press, 1941.

A biography of Emily Brontë appears in Wuthering Heights, *HRW Library Edition. You may wish to share this additional biographical information with your students.*

The poet and novelist Emily Jane Brontë was born on July 30, 1818, the fifth of the six children of Maria and Patrick Brontë. In 1820 the family moved to Haworth in West Yorkshire, where Mr. Brontë remained as rector until his death in 1861. The parsonage where Emily spent all but a few months of her life sat on the edge of the bleak and windy moors and was bounded on three sides by a graveyard. Within a year and a half of moving to Haworth, Mrs. Brontë died of cancer, and her sister, Elizabeth Branwell, joined the household to care for the children. Emily's aunt disliked the harsh Yorkshire weather. Afraid of catching cold, she stayed in her bedroom most of the time. The children were often left on their own, like Catherine and Heathcliff in *Wuthering Heights,* to play on the moors. In addition to performing his pastoral duties, Mr. Brontë, who himself had literary inclinations, tutored his children. They read voraciously and entertained one another with elaborate and inventive stories. Their isolation from the world promoted strong bonds but made them abnormally shy. Yet out of this strange, sequestered environment emerged three brilliant novelists—Emily and her sisters Charlotte and Anne.

In 1824, Mr. Brontë sent his older daughters, Maria and Elizabeth, to the Clergy Daughters' School at Cowan Bridge in Lancashire. Emily and Charlotte soon joined them. The school's fees were low, but the food was inadequate and unappetizing, and the discipline was harsh. Weakened by the poor conditions, the two elder daughters fell sick and died of tuberculosis the following year. Charlotte would use her memories of this tragic period in *Jane Eyre.*

Emily and Charlotte returned home at the end of the school year. Along with Anne and their brother Branwell, they entertained themselves by inventing the imaginary worlds of Angria and Gondal. They created complex romantic tales about these lands and recorded them in miniature books, written in tiny script. When Charlotte went away to Roe Head to teach, Emily and Anne focused their energies exclusively on the chronicles of Gondal. Though the prose is lost, many scholars believe Emily's poems had their origins in the Gondal sagas.

In 1832, Charlotte returned from school and tutored Emily and Anne. When Charlotte returned to Roe Head in 1835, she took Emily along with her as a pupil. Emily grew homesick whenever she was away from Haworth and soon returned. She left the parsonage again in 1837 to take a position as a teacher at Law Hill near

About the Writer (cont.) — *Wuthering Heights*

Halifax, but she returned home again in six months. Despite Emily's tendency toward homesickness, in 1842, hoping to establish a school of their own, Emily and Charlotte went to Brussels to improve their French and German. The sisters were summoned home when their aunt died in October, and Charlotte eventually returned to Brussels alone; Emily chose to remain at home.

When Charlotte came back to Haworth, she advertised for pupils, but none responded. Then Branwell was dismissed from his job and returned home. The sisters abandoned their efforts to open a school at the parsonage in order to care for him. Branwell incurred debts and grew increasingly addicted to alcohol and opium. Critics believe that Emily partially modeled the characters of both Hindley Earnshaw and Heathcliff in *Wuthering Heights* after her brother.

Emily had been secretly writing poems, and when Charlotte discovered them, the three sisters began another scheme to earn a living. They collaborated on a book of poetry published at their own expense in 1846 as *Poems by Currer, Ellis and Acton Bell*. According to Charlotte, the sisters chose masculine pseudonyms because they were "adverse to personal publicity," but more importantly because they "had a vague impression that authoresses are liable to be looked on with prejudice" by critics.

As Branwell declined, spending his time either drunk, writing mournful poetry, or drawing horrible sketches of himself being consumed by an inferno, Emily and her sisters struggled to take care of him. Meanwhile, all three sisters began to work on novels during the day and read aloud their new chapters around the fireplace in the evening. Charlotte complained that scenes from *Wuthering Heights* "banished sleep by night and disturbed mental peace by day."

Though their book of poetry sold only two copies and received few reviews, the experience encouraged the sisters to try to find publishers for their novels. The next year, all three sisters succeeded. However, the publisher who accepted the novels, Thomas Newby, did so only on the condition that the authors advance fifty pounds, to be refunded when two hundred and fifty copies of the book were sold. Newby proved to be an unsavory businessman. He delayed publication for months, ignoring letters from the Brontës and never making the corrections they had requested. It was not until Charlotte's (Currer Bell's) *Jane Eyre* was published in December 1847 and began to sell well that Newby resumed printing Anne's (Acton Bell's) *Agnes Grey* and Emily's (Ellis Bell's) *Wuthering Heights*. As a result, the two novels came out a few months after *Jane Eyre* was published.

The timing was unfortunate. Many critics judged the books in comparison with one another, not on their own merits, and many readers, assuming that all three books were written by the same author, suspected fraud. Newby, hoping to cash in on the soaring sales of *Jane Eyre*, encouraged this suspicion. He never refunded any money to the Brontës.

Emily's disappointment was followed by tragedy. On September 24, 1848, Branwell died. Emily caught a cold at his funeral, failed rapidly, and died of tuberculosis on December 19. She was thirty years old.

About the Novel — *Wuthering Heights*

Special Considerations

Possible sensitive issues in this novel are the death of parents, elopement, a rigid and religiously judgmental character, an abusive and violent household, ethnic bigotry, attempted murder, marriage of first cousins, and the belief of some characters that death is a noble way to escape problems.

For Listening

Wuthering Heights. The Audio Partners, 1998. A dramatization narrated by Patricia Routledge.

Wuthering Heights. Bantam Books, 1996. A BBC Radio Presents audio performance of the novel.

A Study Guide to Emily Brontë's Wuthering Heights. Time Warner, 1994. A presentation narrated by Michael York.

Wuthering Heights. Naxos of America, Inc., 1995. A dramatization on three CDs.

For Viewing

Wuthering Heights. 1939, black and white, NR. Directed by William Wyler, this production stars Laurence Olivier and Merle Oberon and was nominated for eight Academy Awards including best film, best director, best actor, best actress, and best screenplay. Though the action of the film stops at Chapter XVII of the novel, this adaptation is heralded for its characterization of Heathcliff and Cathy and its portrayal of their relationship.

Emily Brontë's Wuthering Heights. Paramount, 1992, PG. This is the first film produced by Paramount's European studios; it was directed by Peter Kosminsky and stars Ralph Fiennes and Juliette Binoche. Fiennes plays an intense and tortured Heathcliff to Binoche's portrayal of both Cathy and her daughter.

Wuthering Heights. Mobil Masterpiece Theatre, 1997, NR. A made-for-television adaptation of the novel starring Orla Brady and Robert Cavanah.

Literary Context

Emily Brontë straddled two eras of British literary history. Though *Wuthering Heights* was actually written and published in the second decade of the High Victorian Period (1830–1880), Brontë's poetry and this single novel seem to distill the essence of the period that preceded it, the Romantic Period (1780–1830). *Wuthering Heights* is written in a dramatic tone and lyrical style. The main male character, Heathcliff, can be seen as a Byronic hero (see below). The setting is rural, and local dialect is used—all elements of Romantic literature.

The Romantics wrote in reaction to the earlier emphasis in Britain on order, rational thought, and science. They challenged previously held beliefs that a powerful government and church led to a harmonious society and that city life was superior to natural splendor. The Romantics believed that spiritual truth could be found in nature (as well as in organized religion), and that truth was best expressed in feelings, emotions, and imagination. The Romantics focused on rural settings and lives of simplicity.

These Romantic writers were also reacting to political and social turmoil. Britain was rapidly changing from an agrarian to an industrial society. The American Revolution caused the defeat of British troops in 1781, and the violent French Revolution led to the rise of Napoleon. This chaos abroad resulted in a more repressive and rigid British government at home.

William Blake, the first published Romantic poet, drew his inspiration from the Bible, Shakespeare, and radical political tracts. He supported both the American and the French revolutions and opposed the subordination of women.

The poets William Wordsworth and Samuel Taylor Coleridge were contemporaries and friends. Wordsworth's poetry often depicted common people in rustic settings; he wrote of basic relationships in an intense emotional and lyrical style, while Coleridge focused on the supernatural.

In Scotland, which is nearer to Yorkshire than Yorkshire is to London, Robert Burns was writing lyric poetry in dialect (focusing on the common man), and many of his topics also concerned the supernatural. Sir Walter Scott, another lyric poet, father of the historical novel and one of two commercially successful poets during the Romantic period, also lived in Scotland.

The other commercially successful lyric poet was the handsome, flamboyant George Gordon, Lord Byron. He created the Byronic

About the Novel (cont.) — *Wuthering Heights*

hero, a proud, moody, cynical, defiant, implacable man, often seeking revenge and often filled with a deep and strong passion for his heroine. Like his contemporaries, Byron wrote of political events. He was a friend of Percy Bysshe Shelley. The intensity of one of Shelley's poems, "Alastor," has been compared to that of Catherine's final embrace with Heathcliff.

The wild, mysterious, natural elements of *Wuthering Heights* define it also as a Gothic novel, a genre that was popular during both the Romantic and the Victorian periods. Percy Bysshe Shelley's wife Mary created the scientific Gothic style at the age of nineteen when she wrote *Frankenstein*. Around the time *Wuthering Heights* appeared, novels with Gothic themes were published by Charles Dickens (*The Haunted Man*) and Hans Christian Andersen (*The Shadow*). Four decades later, Robert Louis Stevenson wrote *The Strange Case of Dr. Jekyll and Mr. Hyde*. Gothic fiction remains popular to this day.

Although *Wuthering Heights* was initially overshadowed by *Jane Eyre*, written by Emily's sister Charlotte, both novels are now considered classics of British literature. Emily, Charlotte, and Anne are honored in the Poets' Corner of Westminster Abbey, England's highest tribute to its writers.

Critical Responses

Wuthering Heights is now widely regarded as one of the greatest English novels, and many critics today contend that Emily's achievement surpasses that of her sisters Charlotte and Anne. Early reviews, however, were mixed. In 1847, the reviewer for the *Athenaeum* found the story "disagreeable." Most early reviewers objected to the novel's subject and characters, but some were willing to acknowledge the author's powerful execution. *Spectator*, for example, claimed that the novel's

> incidents and persons are too coarse and disagreeable to be attractive, the very best being improbable with a moral taint about them, and the villainy not leading to results sufficient to justify the elaborate pains taken in depicting it. The execution, however, is good: grant the writer all that is requisite as regards matter, and the delineation is forceful and truthful.
>
> —an anonymous review in *Spectator* (December 18, 1847)

The reviewer for *Atlas* (XXIII, 1848) said that the novel's "general effect is inexpressibly painful." Nevertheless, this reviewer found "evidences in every chapter of a sort of rugged power."

Tait's Edinburgh Magazine (XV, 1848) complained that the novel lacked sufficient moralizing and failed to teach "mankind to avoid one course and take another."

Later Victorians generally acknowledged the novel's stylistic power while dismissing the book as, for instance, "a nightmare of the superheated imagination." Sir Leslie Stephen, the first editor (1882–1890) of the monumental *Dictionary of National Biography*, did not think Emily worthy of a separate entry and included only a brief discussion of *Wuthering Heights* under Charlotte's entry:

> The novel missed popularity by general painfulness of situation, by clumsiness of construction, and by absence of astonishing power of realization manifest in *Jane Eyre*. In point of style it is superior, but it is the nightmare of a recluse, not a direct representation of facts seen by a genius.
>
> —*Dictionary of National Biography*

In *The Brontë Family* (1886), Francis Leyland even tried to prove that Emily was not the novel's creator, ascribing authorship instead to Branwell. As late as 1913, George Edward Bateman Saintsbury dismissed *Wuthering Heights* as "one of those isolated books,

About the Novel (cont.)   *Wuthering Heights*

which whatever their merit, are rather ornaments than essential parts in novel history."

By the 1880s however, the critical cloud overshadowing *Wuthering Heights* had begun to dissipate. As an anonymous writer in the American journal *Atlantic Monthly* astutely observed, the book

> was curiously and, so far as its author was concerned, distressingly in advance of its time. . . . The utmost which this grim tale found, for years, at the hands even of its most merciful critics was apology; then a generation arose for which it possessed a sort of fearful fascination; and now, at last, it commands a cult, and is acknowledged to have founded a school.
> —"Girl Novelists of the Time," *Atlantic Monthly* (LX, 1887)

In the twentieth century, the novel was the subject of hundreds of critical studies exploring such issues as the book's themes; its narrative construction and chronology; Brontë's sources for the setting, characters, and plot; comparisons to Elizabethan drama and to other European literatures; and autobiographical and psychological interpretations. Controversy continues to surround the novel.

> If *Wuthering Heights* adjusts the conventional paraphernalia of the Gothic, its unquiet graves, its explosive passion, its illicit relationships, its wild landscapes, and its tempestuous climactic conditions, it remolds them into an unconventional shape that neither follows nor creates precedents. Despite its utterly assured mastery of form, it remains the most unconventional and demanding of all English novels.
> —Andrew Sanders, *The Short Oxford History of English Literature*

Critics still attempt to pinpoint *Wuthering Heights* as both a Victorian novel and a Romantic one:

> Without question the most confident celebration of the individual will and passions in English fiction is Emily Brontë's *Wuthering Heights,* a novel whose action is deliberately set far away from the populous haunts of men, and in which such social conventions, duties and restraints as are allowed room appear only in order to be swept away in a tempest of uncurbed emotion. In this, the novel is exceptional in Victorian fiction, where one of the principal concerns is usually the relationship—complex or strained though it may be—of the individual with his or her society.
> —David Skilton, *The English Novel: Defoe to the Victorians*

On the basis of *Wuthering Heights,* Emily Brontë may be called the Romantic novelist par excellence, shunning, like Shelley or Byron, "the common paths that others run" (as she writes in one of her poems) and casting "the world away" for a "God of visions." She was far more uncompromisingly Romantic than her sisters; her credo might be taken from her poem "To Imagination":
> So hopeless is the world without
> The world within I doubly prize
> —Donald F. Stone, *The Romantic Impulse in Victorian Fiction*

As Melvin R. Watson notes in his survey of the novel's critical history,

> Never, probably, will an interpretation of *Wuthering Heights* be made which will satisfy all people for all time, for a masterpiece of art has a life all its own which changes, develops, and unfolds as the generations pass. . . . Like the wuthering heights themselves, it has withstood the fiercest tempests; it stands sturdy yet, a monument of proof that a work of art can be strengthened by the sunny breezes of favorable criticism, but cannot be destroyed by critical storms.
> —Melvin R. Watson, "*Wuthering Heights* and the Critics" in *Nineteenth-Century Fiction* (III, 1949)

The Novel at a Glance — *Wuthering Heights*

Plot and Setting
Wuthering Heights is the story of the intense love between Catherine and Heathcliff that affects two families spanning three generations. The action takes place in 1771–1802 at Wuthering Heights, the family home of the Earnshaws, and at Thrushcross Grange, the family home of the Lintons—both on the heath of the northern English county of Yorkshire.

A **Novel Review** that includes **plot** and **setting** appears on page 63 of this Study Guide.

Structure and Point of View
Structure and point of view are complex and interwoven. The story begins on a winter day in 1801 and closes on a September day of the following year. Within this framework is a story spanning thirty years, told in flashbacks. While the voice is first person throughout, there are layers of shifting narrators: The novel purports to be the diary of Mr. Lockwood. Early on, Brontë complicates this first-person point of view by introducing Nelly Dean, whose words are quoted verbatim or "a little condensed" by Lockwood. Within Nelly's narration, other narrators emerge briefly to describe events at which Nelly was not present: Heathcliff, Isabella, Cathy, Zillah, Linton.

A **Literary Elements Worksheet** that focuses on **Multiple Narrators** appears on page 65 of this Study Guide.

Major Characters
Heathcliff is a dark-featured gypsy foundling brought to live at Wuthering Heights by its owner Mr. Earnshaw, father of Hindley and Catherine. Mr. Earnshaw's preference for Heathcliff over his own son leads to tragic consequences. An assault on Heathcliff's dignity at the hands of the jealous Hindley and the loss of his true love Catherine to Edgar Linton drive him to plan vengeance that dominates his life and relationships.

Catherine Earnshaw is a wild, passionate beauty who roams the moors with Heathcliff as a child and considers him her soulmate, yet marries her neighbor Edgar Linton for his wealth and social position. Her death at eighteen breaks the hearts of the two men who love her.

Edgar Linton, the master of Thrushcross Grange, is as blond, even-tempered and kind as his wife is dark, fiery, and manipulative. He becomes a caring father who adores their daughter, Cathy, but even he cannot protect her from Heathcliff.

Isabella Linton, the pale, immature younger sister of Edgar, is disowned by him after she elopes with Heathcliff, who sees her as an instrument of revenge.

The Novel at a Glance (cont.)   *Wuthering Heights*

Hindley Earnshaw lives a life scarred by loss. First, the loss of his father's favor to Heathcliff results in destructive jealously; then, the loss of his wife Frances sets him on a path of gambling and excessive drinking. Ultimately, these result in his forfeiting both Wuthering Heights and his son to Heathcliff.

Catherine Linton, known as Cathy, inherits her mother's beauty and vivacious spirit and her father's generous nature. By her naiveté she falls prey to Heathcliff's plan of revenge, but by her strength of character she overcomes it.

Hareton Earnshaw is the handsome and proud, yet socially inept and uneducated, son of Hindley. By denying Hareton opportunity and education, Heathcliff takes revenge for the treatment he received from Hareton's father Hindley.

Linton Heathcliff is the pale, sickly son of Isabella and Heathcliff. He also becomes an instrument of his father's revenge.

Nelly Dean is a servant whose mother also worked at Wuthering Heights. Nelly was a playmate of Catherine and Hindley and witnessed Heathcliff's arrival. She is confidante to Catherine, Heathcliff, and Cathy, and therefore an effective narrator of their story.

Theme

The novel explores the themes of **love as a creative and nurturing force versus love as an all-consuming and destructive force.** The **destructive power of revenge** and the **consequences of passion that dominates reason** are also key themes.

Symbolism

The houses: Wuthering Heights sits on a desolate and stormy hill surrounded by gnarled, stunted trees. Thrushcross Grange is calm, ordered and peaceful. It is elegantly decorated with carefully tended gardens. The houses represent the emotions and family structure of the Earnshaws (and Heathcliff) and the Lintons.

Natural elements: The weather, the plants around the houses, and the environment of the moors often reflect and even **foreshadow** the emotions or conflicts of the characters.

Style

The **Gothic** style popular in the late eighteenth and early nineteenth centuries is characterized by gloomy settings and an atmosphere of terror and mystery. *Wuthering Heights'* setting, ghostly apparitions, melodramatically passionate characters, and fascination with the spiritual union of souls embody this Gothic style.

A **Literary Elements Worksheet** that focuses on **Symbolism** appears on page 64 of this Study Guide.

A **Literary Elements Worksheet** that focuses on **Gothic Literary Elements** appears on page 68 of this Study Guide.

Introducing the Novel   *Wuthering Heights*

Options

Engaging Issues

The central issue of love in Wuthering Heights *continues to be both relevant and controversial. Use this activity to engage students in grappling with traditional and romanticized notions of love.*

- Prepare survey for students about love, romance, and marriage. Ask students to label each statement as *true* or *false*.

Possible sample statements

1. Love is painful.
2. Everyone has a soulmate.
3. Love is an emotion.
4. Love is a decision.
5. All is fair in love.
6. Jealousy is an expression of love.
7. When you fall in love, it should be forever.
8. To die for one's love is noble.
9. You should marry only your soulmate.

- Small groups choose a statement and discuss both sides, giving reasons for their opinions. Each group reaches consensus on the dilemma and on the broader issue and presents its conclusions to the class. How do the conclusions compare with the survey responses?

FILM CLIP

Inferring/Predicting

Choose a short segment from one of the film versions of *Wuthering Heights* to show the class. Ask students
- What is the relationship of these people?
- What has happened to cause this situation?
- Whose side would you take?
- What will happen next?
- What are three other questions you expect the book to answer?

VISUAL PRESENTATION

"Flung . . . out into the middle of the heath"

Introduce the novel's settings: the heath that both Catherines love and the homes in which they live. Provide a group of students with travel brochures, calendars, illustrated architecture books, or magazines featuring the Yorkshire area of England and houses of the era of the novel. Pre-select descriptive passages from *Wuthering Heights*, give them to group members, and have students locate pictures representing that which is described. Students should present pictures accompanied by an expressive reading of the passage to their classmates.

AUDIOVISUAL INTRODUCTION

Video Views

- *The Brontë Connection* traces the discovery of the personal papers of the Heatons, a family of Yorkshire gentry, which identify the real-life Heathcliff. A 53-minute color film, it is available through Films for the Humanities & Sciences.
- In *Wuthering Heights: A Critical Guide*, imagery and narrative style are examined. A 52-minute color film, it is available through Films for the Humanities & Sciences.
- To check other offerings, log on to **http://www.films.com**

Plot Synopsis and Literary Elements

Wuthering Heights

Chapters I–III

Plot Synopsis

It is the winter of 1801 in Yorkshire, a county in northeastern England. Mr. Lockwood, the new tenant of Thrushcross Grange, has just returned from his first visit to Wuthering Heights to meet his landlord, Mr. Heathcliff. Lockwood initially considers Heathcliff to be a kindred spirit: another loner who is cynical, reserved, and suspicious. Because his own natural reticence the previous year had cost him the love of a beautiful woman, Lockwood is more understanding than offended at Heathcliff's refusal to shake hands. He is intrigued to be renting from this morose man whose appearance is in contrast to his gentlemanly dress and comfortably furnished home. Lockwood's interest turns to distress when his host's dogs become vicious. Lockwood is ready to depart in a huff until soothed by Heathcliff, who has returned with wine. Their subsequent pleasant, intelligent conversation prompts Lockwood to announce he will return the following day to Wuthering Heights.

The next day Lockwood walks the four miles "through heath and mud" to Wuthering Heights atop its "bleak hill-top." By the time he arrives, it has begun to snow. Despite the weather, the servant Joseph rebuffs his knock, suggesting that he head for the fields if he wants to talk to Heathcliff. Finally, Lockwood is saved from the cold by a young, coatless man carrying a pitchfork who leads him inside. The interior is warm and pleasant but the inhabitants are not. The young, pretty woman Joseph had referred to as the "missis" is snappish and scornful. The young man Lockwood believed to be a servant does not attend to the lady and looks upon him with disdain. Heathcliff's arrival temporarily puts his visitor at ease, but his ferocious demands on the young woman to provide tea and his rude corrections of Lockwood's honest mistakes break that spell. Lockwood learns the young woman is Heathcliff's daughter-in-law. She is a widow, and both Heathcliff and Joseph seem to enjoy tormenting her. The young man introduces himself defiantly as Hareton Earnshaw, not her husband.

Heathcliff refuses Lockwood's request for a guide back to Thrushcross Grange, and, during an unpleasant dinner, Heathcliff displays open hostility toward both young people. Lockwood changes his good opinion of his host and privately asks Mrs. Heathcliff for help—only to discover that she is not allowed to leave the house. He resigns himself to spending the night until Heathcliff informs him that he must sleep with Joseph or Hareton. Insulted, he tries to leave, but the dogs knock him down, and Heathcliff and Hareton laugh at the spectacle. Zillah the housekeeper takes Lockwood inside, and left with little choice, Lockwood decides to spend the night.

Zillah leads Lockwood to a bedroom Heathcliff doesn't want anyone to use, though she can't explain why. Finally alone, Lockwood realizes the bed is hidden behind panels that form a private closet with a small window. Scratched repeatedly in the window ledge paint are the names Catherine Earnshaw, Catherine Heathcliff, and Catherine Linton. He discovers a Bible with *Catherine Earnshaw* and a date from almost a quarter century past written on the flyleaf. Catherine has used the margins as a diary. Lockwood reads how her brother Hindley, after their father's death, mistreated Heathcliff; how he forced her and Heathcliff to sit through Joseph's long-winded sermons; and how Hindley finally barred Heathcliff from contact with the family. Lockwood falls into a sleep tormented by nightmares.

In the first nightmare, Lockwood sits through an endless sermon on sin that deteriorates into a violent riot among the members of the congregation. Lockwood awakes, but soon drifts off again. In his second nightmare, in order to stop a branch tapping on his window, he breaks the glass, reaches out, and touches a small, cold hand. A melancholy voice—identifying itself as Catherine Linton—begs him to let

Plot Synopsis and Literary Elements (cont.)

Wuthering Heights

her in. Lockwood sees a child's face and, desperate to escape the hand that now grasps him, vainly rubs its wrist on the broken pane until blood soaks the sheets. Only when he promises to admit the apparition does it loosen its grip. He then piles books against the pane, but the voice continues to plead that it has been homeless for twenty years. When the books seem to move, Lockwood cries out, and Heathcliff rushes into the bedroom. Heathcliff is agitated, and Lockwood claims the room is haunted.

After Lockwood explains how he spent his time before falling asleep, Heathcliff becomes noticeably emotional and asks Lockwood to leave the room. Before doing so, Lockwood witnesses what he calls his landlord's superstition. Heathcliff gets on the bed, wrenches open the window, and in an anguished voice begs Catherine to come in.

At the first sign of dawn Lockwood departs, with Heathcliff guiding him to the entrance of Thrushcross park. Upon reaching the house, Lockwood is greeted by the servants, who feared he had died in the snowstorm.

Literary Elements

Point of View: The narrator, Mr. Lockwood, like the reader, is trying to piece together the history of Wuthering Heights. Through this first-person narrative style, Brontë establishes an atmosphere of mystery and suspense.

Symbolism: "'Wuthering' being a significant provincial adjective, descriptive of the atmospheric tumult to which its station is exposed in stormy weather." Weather and property are central symbols for relationships and personalities in the novel. The interior climate Lockwood finds at Wuthering Heights mirrors the storm outside. Its situation on a "bleak hilltop" of hard earth reflects the circumstances of the inhabitants of the house and the hard heart of Heathcliff. Heathcliff's name is symbolic of his harsh and dangerous personality.

Characterization: Lockwood may be a slightly biased first-person narrator, but the reader is provided with **direct characterization** for the appearance and personality of the central characters through his observations. His recounting of actions and conversations, **indirect characterization,** gradually reveals the dynamic of the relationships of these characters. Heathcliff, Mrs. Heathcliff, Hareton, and Lockwood are **dynamic characters.** Their unpredictable behavior reflects the multiple dimensions of their personalities.

- **Lockwood,** a gentleman from London, "tolerably attractive," with money that enables him to rent the magnificent Thrushcross Grange, is intrigued by his landlord, whose reserve makes him feel like an extrovert.
- **Heathcliff** is "dark-skinned," "handsome," and "morose." His exchanges with Mrs. Heathcliff and Lockwood are as stormy and cold as the winter weather outside the manor house. Yet, factors that could have shaped the hardhearted Heathcliff are unveiled when the reader learns from Catherine Earnshaw's diary that he was mistreated as a child. His "uncontrollable passion of tears" while begging for the return of Cathy's ghost reveals the overwhelming emotion of which he is capable.
- **Mrs. Heathcliff** is slender, with "the most exquisite little face . . . small features, very fair; flaxen ringlets." She says little and regards Lockwood in a cool, scornful manner. She is obviously unhappy and picks a fight with the servant, whom she calls a "hypocrite" and threatens to harm with a witch's spell. However, she does meekly speak up for the narrator, recommending someone lead him home so that he will not be lost in the storm. The reaction this suggestion elicits from Hareton and Joseph indicates her oppression in the house.
- **Joseph,** the servant, is "vinegar" of face and manner. His Yorkshire dialect is peppered with religious references, which hint at his religious zealotry that Mrs. Heathcliff describes as hypocritical.

Plot Synopsis and Literary Elements (cont.)

Wuthering Heights

- **Hareton** appears first to be a servant and exhibits pride that to Lockwood seems misplaced because he is shabbily dressed and appears uneducated. Lockwood observes he "looked down on me, from the corner of his eyes, for all the world as if there were some mortal feud unavenged between us."

Style
Gothic elements
- secluded, wind-swept locale
- dark house with morose, violent host
- dream of a mysterious ghostly child

Foreshadowing
- The argument between Joseph and Mrs. Heathcliff hints at what the reader will soon come to know about Catherine Earnshaw and Joseph's feelings toward her: "[Y]ou'll never mend your ill ways, but go right to the devil, like your mother before you!"
- Heathcliff steps on a book titled "The Broad Way to Destruction." His actions **symbolize** and **foreshadow** his future tragedy.
- The first dream that Lockwood has is one of a sermon on forgiveness of one's brother seventy times seven. Lack of forgiveness between brothers will set off the chain of events that dooms Heathcliff.

Theme: **Passion overcoming reason** is illustrated by Heathcliff's entry into Lockwood's room and his sobbing for the apparition to return. In addition, groundwork is laid for the comparison of **love as a nurturing and creative force versus love as an all-consuming and destructive force.**

Chapters IV–IX

Plot Synopsis

That night, still in bed recovering from his ordeal, Lockwood's housekeeper, Nelly Dean, tells him that Heathcliff is rich and greedy, admitting that Heathcliff could live in the Grange, which is a finer house, but prefers to collect the rent money for it. Young Mrs. Heathcliff, Catherine Linton, grew up at the Grange; she is the daughter of Mrs. Dean's late master, Edgar Linton. Linton's sister married Heathcliff; their son married young Catherine Linton. Hareton Earnshaw is Catherine Linton's cousin and the rightful heir to the property. He is unaware that Heathcliff has cheated him out of it.

Nelly Dean then takes the story back to the fateful summer day when Mr. Earnshaw, the master, brought home a boy who had been abandoned to starve on the Liverpool streets. They named him Heathcliff, after a son who died in childhood, but did not give him the Earnshaw surname. Heathcliff was soon Mr. Earnshaw's favorite, much to the distress of Mrs. Earnshaw and Hindley, who bullied him. Heathcliff and Catherine, however, became best friends. By the time Mrs. Earnshaw died, two years later, the boys actively hated each other. Heathcliff, a sullen and hard-natured boy, is fully aware of his position as Mr. Earnshaw's favorite and coolly takes advantage of his status to blackmail Hindley and take possession of his colt. Nelly admits that she failed to realize then how vindictive Heathcliff could be.

As Mr. Earnshaw's health begins to fail, he becomes less tolerant of family conflicts, especially criticism of Heathcliff. Increasingly aware of the stress on the old man, the curate, who teaches the children, advises him to send Hindley to college and Earnshaw does. This does not bring peace because the sickly Mr. Earnshaw is easily manipulated by Joseph, who worries him about the state of his soul and counsels that children should be reared more strictly. Joseph is most critical of Catherine. Her father's harsh words sometimes hurt Catherine, but she is more able to laugh and forgive, while Heathcliff broods.

On a blustery fall evening, with Catherine and Heathcliff at his knee, Mr. Earnshaw dies peacefully in

Plot Synopsis and Literary Elements (cont.)

Wuthering Heights

his sleep. When Hindley returns for the funeral after an absence of three years, he is accompanied by a wife, Frances. He confines Nelly's and Joseph's living quarters to the back-kitchen and treats Heathcliff as an ordinary farm laborer. When the boy is no longer permitted instruction from the curate, Catherine secretly teaches Heathcliff and accompanies him in the fields. Though Hindley is generally indifferent to the children's behavior, Joseph and the curate reprimand Hindley for his neglect; this reminds him to flog Heathcliff occasionally and deprive Catherine of meals as punishment for missing church.

The children learn to laugh at their oppressors, and their behavior grows more reckless. One rainy evening when the pair are not home for supper, Hindley angrily orders them locked out for the night. Nelly stays up, determined to admit them; hearing footsteps, she runs outside. It is only Heathcliff. Catherine, he tells Nelly, is at Thrushcross Grange recuperating from a dog bite on her ankle. They were spying on the Linton children when the sound of their contemptuous laughter alerted a bulldog that chased after them and bit Catherine. She was carried by a servant into the house, followed by Heathcliff. The Lintons first considered them thieves until their son Edgar remembered Catherine from church. Heathcliff was sent home. The next day Mr. Linton visits the Heights to lecture Hindley about the need to more carefully oversee his sister's conduct. Heathcliff is forbidden to speak ever again to Cathy.

Catherine spends five weeks at the Grange. Frances visits often, bringing new clothes and flattering talk. When Catherine returns to the Heights on Christmas Eve, she is beautifully groomed and displays charming manners. Heathcliff, however, has been completely neglected and is dirty and unkempt. Still, Catherine rushes forward and kisses him. After she, not unkindly, notes he is dirty and unconsciously compares him to the Lintons, he draws back. She does not understand how this has hurt his pride. When Hindley maliciously encourages him to shake hands properly, Heathcliff refuses to be ridiculed and rushes away.

As Nelly prepares for a Christmas visit from the Linton children, Heathcliff finally asks Nelly to help him with his appearance, and he tells her he is going to be good. As she washes and combs his hair, the boy confesses he wants to be fair skinned and rich like Edgar Linton. Nelly encourages him to be less sullen and take pride in himself. When the Lintons arrive, he is eager to present himself, but after being insulted, throws hot applesauce on Edgar. When Hindley takes Heathcliff away to beat him, Catherine scolds the Lintons for interfering and for crying about the incident when they are unhurt. Eventually she seems to relax in their company. During dinner Catherine's distress at Heathcliff's banishment is obvious, and later in the evening she goes to the garret to talk to him through the door. So Hindley will not learn of this forbidden contact, Nelly returns to warn them and finds Catherine is inside the garret. When they come out Catherine asks Nelly to feed Heathcliff, but the boy cannot eat and stares into space. He tells Nelly he is dreaming of ways to exact revenge on Hindley. Nelly counsels forgiveness, but he says that this planning is the only way he can stand the pain. Nelly abruptly interrupts her story to apologize to Lockwood for chattering. He begs her to continue and she skips to the following summer, June 1778.

That summer the Earnshaw heir, Hareton, is born, but Frances dies of consumption soon after giving birth. Hindley wants nothing to do with his son and places the baby in Nelly's care. To assuage his grief, Hindley begins drinking heavily and keeping company with an unsavory crowd. Soon, his conduct has run all servants off except Joseph and Nelly. While Heathcliff takes delight in Hindley's decline, Catherine becomes haughty and headstrong. They remain constant companions, but to Heathcliff's chagrin, Catherine is also the object of Edgar Linton's attention and occasional visits.

In the company of the Lintons, Catherine is cordial and refined, but one afternoon when Edgar is visiting, she strikes Nelly in a fit of pique at her presence as

Plot Synopsis and Literary Elements (cont.)

Wuthering Heights

chaperone, shocks Edgar by violently shaking Hareton, and turns on Edgar when he intervenes to save the child. Edgar is appalled and leaves the house. He returns, however, and when Nelly later enters to warn that a very drunken Hindley is back, she believes the couple have confessed their love.

To avoid Hindley, Edgar quickly leaves, and Catherine retires to her room. In a violent rage, Hindley first tries to attack Nelly, then dangles Hareton from the second floor banister. As Heathcliff unwittingly enters below, the baby falls. By a natural impulse, Heathcliff catches the child but becomes anguished when he looks up and realizes that he has "made himself the instrument of thwarting his own revenge" against Hindley.

Nelly takes Hareton to the kitchen to soothe him. She thinks Heathcliff has gone to the barn, but actually he is sitting quietly in another part of the room. Catherine enters and asks about Heathcliff's whereabouts. Assured that he is in the barn, Catherine confides to Nelly that Edgar has asked her to marry him and that she has accepted. She confesses that she has no business marrying Edgar but that because Hindley has brought Heathcliff so low it would degrade her to marry him now. At this point, Nelly spots Heathcliff quietly leaving the kitchen but does not alert Catherine. Heathcliff does not hear Catherine go on to confess her love for him: ". . . he's more myself than I am. Whatever our souls are made of, his and mine are the same; and Linton's is as different as a moonbeam from lightning, or frost from fire." Further, she says that marrying Edgar will permit her to aid Heathcliff and to place him out of her brother's power. Nelly then tells Catherine that Heathcliff overheard much of what was said.

Although a furious thunderstorm has arisen, Catherine stands outside until after midnight awaiting Heathcliff's return. Finally, Catherine comes back inside. Refusing to change her wet clothes, she sits up all night in the kitchen. By morning her teeth are chattering. Joseph tells Hindley she was out all night with Heathcliff, a charge she denies though she does admit to Edgar's secret visits after Joseph reveals them. When Hindley says he will send Heathcliff away, Catherine says she will go with him, then becomes uncontrollably grief-stricken. Her chill develops into a delirious fever. Mrs. Linton visits several times before moving Catherine to the Grange to convalesce. While she recovers, Mr. and Mrs. Linton both catch her fever and die within days. Heathcliff does not return, and three years later she marries Edgar and takes Nelly with her to the Grange.

Suddenly realizing the hour is late, Nelly interrupts her saga so Lockwood can rest.

Literary Elements

Point of View: First-person narration continues with Nelly Dean assuming the role of storyteller. She recounts the saga in **flashback** to Mr. Lockwood. She is admittedly biased. Nelly is generally a sympathetic character, though the reader may not always agree with her actions, which contributed to the situations she is describing.

Conflict
Internal conflict
- Heathcliff seems torn by his desire to please Cathy and his need to alter his appearance or conversation in order to accomplish that. Heathcliff expresses a desire to be decent and good, but he discovers that the only way to protect himself from the harsh treatment of others is to allow his more malicious nature to flourish.
- Catherine's conflict is between two opposing aspects of her personality. This internal conflict is manifested in her relations with Heathcliff and Edgar. Each seems an embodiment of one side of her personality—Heathcliff as unbridled nature; Edgar as civilized behavior (see **symbolism**).
- Hindley's drunken ravings and his relief at Hareton's rescue reveal his love of the child, but his grief and inability to express that love have created a tyrant.

Plot Synopsis and Literary Elements (cont.)

Wuthering Heights

External conflict

- Heathcliff and Hindley have been at odds since Heathcliff's arrival. Hindley, feeling supplanted by Heathcliff as the object of his father's affections, systematically degrades Heathcliff by bullying, beating, and humiliating him. Heathcliff swears revenge. Catherine identifies Hindley's degradation of Heathcliff as her reason for not marrying him. Overhearing this admission provides Heathcliff with his most potent motive for destroying Hindley.
- Heathcliff's conflict with Edgar Linton is twofold: he envies Edgar's physical appearance, which seems automatically to earn Linton acceptability in society and in Cathy's heart, and he is jealous of the time Linton spends with Cathy, which evolves into a marriage proposal.

Characterization:

- Despite Nelly's admitted contempt for **Catherine Earnshaw,** the positive aspects of Catherine's personality are also revealed. She is loyal to childhood companions, especially Heathcliff. The strength of their bond is alluded to as they comfort each other after the loss of Mr. Earnshaw. The strength of their relationship is directly stated by Catherine: "[w]hatever our souls are made of, his and mine are the same. . . ."
- **Edgar** and **Isabella** are painted as high-strung and pampered. During the Christmas visit to the Heights, they are not merely civilized; they are overrefined.
- **Heathcliff** is rough from an early life on the streets; this is compounded as Hindley increasingly isolates him. Nelly realizes she underestimated the effect of this: "I really thought him not vindictive: I was deceived completely." The nature of his personality is most starkly visible in juxtaposition to Linton, his **foil.** "The contrast resembled what you see in exchanging a bleak, hilly, coal country for a beautiful fertile valley. . . ." Catherine reiterates this comparison. Her similes for Edgar imply that Heathcliff is his opposite:
  - "Linton's [soul] is as different as a moonbeam from lightning, or frost from fire."
  - "My love for Linton is like the foliage in the woods: time will change it, . . . as winter changes the trees. My love for Heathcliff resembles the eternal rocks beneath."

Symbolism: Though Catherine chooses to marry Edgar, she has not resolved the internal conflict of her divided nature. The storm that rends the **tree** on the night of Heathcliff's disappearance is a vivid symbol for the turmoil splitting her personality.

The **houses** are indicative of the families. Thrushcross Grange is "a splendid place," the household is calm and orderly, the gardens beautiful and neat, and the parents thoughtful and conscientious. The Heights is fraught with turmoil, most of the servants have left, Joseph's religious zeal stirs up bad blood among the inhabitants, and Hindley keeps no control over the children and cares little for them.

Foreshadowing

- The introduction of the character Frances mentions her fear of dying, foreshadowing that event.
- As Nelly watches Edgar reenter the parlor to reconcile with Catherine, her comment foreshadows the nature of their relationship: "He's doomed, and flies to his fate!"

Theme

Love as a nurturing and creative force versus love as an all-consuming and destructive force

- Hindley's love for Frances leads him in grief to forsake his son and destroy himself with alcohol.
- Edgar's response to Catherine's outburst becomes compassionate and nurturing, leading to a profession of love.

Plot Synopsis and Literary Elements (cont.)

Passion over reason
- Hindley's passionate hatred of Heathcliff makes him tyrannical.
- Catherine's emotional nature and passionate belief in a spiritual union with Heathcliff **ironically** leads her to what in her mind is a reasonable decision—marrying Edgar.
- Heathcliff regrets saving young Hareton because it prevents his revenge.
- Heathcliff abruptly leaves the Heights rather than asking Catherine about her decision.

Chapters X–XVII

Plot Synopsis

After four weeks Lockwood is still recovering. Heathcliff, whom he holds partly responsible for his illness, has sent grouse and visited once. Lockwood asks Nelly to continue the saga.

She begins by remembering the first six months of Edgar and Catherine's marriage as relatively smooth. Edgar and his sister Isabella give in to Catherine's wishes because Catherine still suffers from occasional depression, a problem since her earlier illness after Heathcliff vanished. The couple seem increasingly content until a late September day when Nelly finds Heathcliff lurking in their garden. At first she does not recognize the handsome, militarily erect man with a full beard and foreign-sounding voice. Heathcliff commands Nelly to arrange for him to meet with Catherine without saying who awaits her. With misgivings she complies. Cathy is thrilled by Heathcliff's return and dramatically rejects Edgar's suggestion that she and her visitor talk in the kitchen. The parlor is hastily arranged for company and Isabella joins them, making a foursome. Edgar's distress grows at his wife's obvious delight at Heathcliff's return and at their somewhat intimate conversation.

Late that evening, too excited to sleep, Catherine wakes Nelly to discuss Heathcliff's return. Catherine explains to Nelly that Heathcliff came to the Heights to learn her whereabouts and is staying there. Hindley has invited Heathcliff there to provide himself with the opportunity to recoup the money he lost playing cards with Heathcliff. Congratulating herself on being in a position to save her brother from Heathcliff, Catherine tries to persuade Edgar that her friendship with Heathcliff is essential and benign.

Isabella, who is eighteen but immature, becomes infatuated with Heathcliff. Edgar is appalled, knowing that if he and Catherine do not produce a son and Isabella marries Heathcliff, the Linton fortune could pass to Heathcliff. Catherine reveals Isabella's infatuation to Heathcliff when he next visits. After a mortified Isabella flees in tears, he denies any interest in her. Heathcliff's questions to Catherine about who could inherit Thrushcross Grange cause Nelly to wonder where this idea will lead. Nelly suspects the worst.

Suddenly recognizing the active threat that Heathcliff represents to her former master, Hindley, Nelly visits the Heights to warn him. His five-year-old son Hareton not only does not recognize his former nurse but curses her and throws a rock at her. With bribes of oranges, Nelly is able to learn from the boy that Hindley cannot endure his presence. Just as Hindley did to him, Heathcliff has dismissed Hareton's teacher, guaranteeing future disadvantages. Nelly sends Hareton to get his father, but when Heathcliff appears instead, Nelly grows frightened and distressed and runs away.

She is equally disturbed by Heathcliff's increased attention to Isabella and speaks to Catherine about a furtive embrace she has seen between the two. Catherine immediately tries to squelch Heathcliff's

Plot Synopsis and Literary Elements *(cont.)*

Wuthering Heights

pursuit. Their quarrel escalates. Attempting to stop it, Nelly alerts Edgar. When Heathcliff ignores Edgar's order to leave, Edgar tells Nelly to call for servants to assist. Catherine calls Edgar a coward who deserves to be beaten by Heathcliff. Heathcliff starts the fight but Edgar lands the first blow, then leaves to gather reinforcements. Catherine implores Heathcliff to go; Heathcliff swears he'll murder Edgar. When Edgar returns with three men wielding bludgeons, Heathcliff smashes open the locked inner door and escapes. Catherine becomes increasingly frenzied. Certain she could have persuaded Heathcliff to leave Isabella alone, she is angry at both men.

Catherine confides in Nelly her plan to frighten Edgar with a fit to make him pay for his outburst and to avoid further recriminations from him. When Edgar returns, he insists that Catherine choose between him and Heathcliff. She screams to be left alone, dashes her head against the sofa, and falls into a stupor. Edgar is terrified until Nelly tells him it is an act. Incensed, Catherine locks herself in her room and refuses all food. Edgar ignores his wife and warns Isabella he will have nothing to do with her again if she continues to encourage Heathcliff.

At the end of three days, Catherine asks Nelly for food and claims she is dying. After eating the toast and tea Nelly brings, Catherine again threatens to starve herself or recover and flee the country because, she claims ambiguously, "he does not love me at all—he would never miss me!" As her anger grows she becomes more frenzied. Delirious with fever, she thrusts open her window and leans out into the cold. She claims to see the candlelit window of her old room at the Heights and says she will return there when she is dead. She begins talking to Heathcliff and says she will not rest in her grave until he is with her. Edgar enters. Horrified at his wife's wan appearance, he scolds Nelly for having kept her worsening condition secret. Nelly announces she will fetch the doctor.

On her way, she discovers Isabella's white dog hanging and almost dead. While releasing the dog, she hears a horse galloping away and wonders who it could be at two in the morning. The doctor advises Edgar that his wife merely needs tranquil surroundings, but he tells Nelly that insanity is likely. In the morning a servant girl returning from town tells Edgar of Isabella's elopement. Edgar severs all relations with his sister and tells Nelly to send her property to her new home.

Heathcliff and Isabella remain away for two months. During that time, Edgar tenderly nurses his wife, and she begins to recover. While the doctor does not foresee a complete recovery, Edgar remains optimistic, even more so when he learns that an heir is expected. Catherine is easily exhausted, depressed, and talks of death.

Edgar does not respond to his sister's brief letter announcing her marriage and hoping for a reconciliation. The day after the newlyweds return to the Heights, Nelly receives a long letter from Isabella describing her unhappy new life: Hareton curses her and threatens to set his dog on her, Joseph pretends not to understand her refined speech, Hindley appears on the verge of madness and has told her he plans to shoot Heathcliff, and Heathcliff terrorizes her with abusive language. He blames Edgar for Catherine's illness and threatens to make Isabella suffer in her brother's place until he can take revenge directly. Isabella ends her letter by expressing hatred for her husband. She asks Nelly to keep secret her terrible existence and to visit her.

Edgar refuses to forgive Isabella but allows Nelly to visit the Heights. There, Nelly finds the once cheerful house dreary and dismal. Isabella is listless and unkempt. In contrast, Heathcliff is oddly friendly and appears happy. Isabella is visibly disappointed when Nelly admits she carries no letter from Edgar. Heathcliff probes for news of Catherine. Nelly reports she is recovering but will likely never be her former self. He asks Nelly to arrange a meeting between him

Plot Synopsis and Literary Elements (cont.)

Wuthering Heights

and Catherine, but she refuses. He admits his existence would be hell without Catherine and denigrates Edgar's love of Catherine. When Isabella tries to defend her brother, Heathcliff ridicules her loyalty and acknowledges his contempt for her. Isabella explains to Nelly that Heathcliff married her to gain power over Edgar but that she will die or see him dead to prevent that. Nelly agrees to deliver a letter to Catherine and arrange a meeting after Heathcliff threatens to keep her at the Heights overnight.

At this point, Nelly returns to the present to ask Lockwood whether what she did was right or wrong. She believes her actions were expedient but still a betrayal of trust. Lockwood takes up the narration again in Chapter XV, but in the words of Nelly Dean because "I don't think I could improve her style."

Nelly secretly gives Heathcliff's letter to Catherine, who is too distracted to understand it. Nelly tells her that Heathcliff is waiting in the garden to see her. Before Nelly can summon him, Heathcliff enters. He passionately embraces Catherine. She tells him that he and Edgar have broken her heart and asks how many years he intends to live after she is dead. He accuses her of being possessed by the devil and says her words will torment him when she is gone. She begs his forgiveness and asks him to hold her, even as Nelly warns that Edgar has returned from church. Before Heathcliff can leave, Catherine faints in his arms. Heathcliff prevents Edgar from attacking him by placing Catherine in Edgar's arms and asking him to help her. He waits in the parlor as Nelly and Edgar work to bring Catherine to semiconsciousness, but she recognizes no one. Heathcliff insists he will remain in the garden overnight.

Catherine gives birth to the daughter she has carried seven months; before morning she herself is dead. A grieving Edgar lies down beside his wife. At dawn Nelly goes to the garden to tell Heathcliff, but he already knows of the death. He asks whether Catherine mentioned his name. When Nelly sympathetically tells him she never regained consciousness, Heathcliff cries out in anguish, saying her ghost will haunt him because he cannot live without his true soul.

Edgar guards the body, while Heathcliff keeps a silent vigil outside. When Edgar finally leaves to sleep, Nelly allows Heathcliff to see Catherine one last time. After he is gone, Nelly discovers he has removed Edgar's blond hair from Catherine's locket and placed a lock of his dark hair inside. Nelly entwines them and places both in the locket. Catherine is not buried in the Linton chapel or with the Earnshaws. She rests instead in a corner of the churchyard.

The day following the funeral, Isabella escapes from the Heights and comes to the Grange. Calling Heathcliff a beast, she removes her wedding ring, smashes it with a poker, and hurls it into the fireplace coals. She describes a dreadful time after Catherine's death: Hindley wanted to attend the funeral but could not stay sober. When Heathcliff was not at the Grange, he was locked away in his room praying loudly to Catherine, "senseless dust and ashes," and cursing God's name. Hindley tried to murder Heathcliff, but Heathcliff wrestled the pistol from him. During a quarrel the next day, Heathcliff hurled a knife at her head, which struck her beneath the ear. While Hindley wrestled him to the floor, she escaped.

Isabella leaves. A son, Linton, is born and they remain in London until her death twelve years later. Edgar never leaves the Grange except to visit his wife's grave. His greatest joy is their daughter, named for her mother. Hindley dies six months after his sister's death, and ownership of the Heights passes to Heathcliff, to whom Hindley has mortgaged all his property. Heathcliff now plans to treat Hareton with the same cruelty that Hindley had shown him.

Literary Elements

Point of View: While Nelly is the primary narrator, Isabella adds to the story from first-person perspective when she describes her arrival at Wuthering

Plot Synopsis and Literary Elements (cont.)

Wuthering Heights

Heights after the honeymoon and again when she describes her escape from the Heights.

Symbolism: The vision of the world, literally, that Edgar and Catherine "gazed on, looked wondrously peaceful" prior to Heathcliff's arrival. From their parlor window, they could see the Thrushcross Grange garden and park, valley, and mist; Wuthering Heights, though, was not even visible from their vantage point, symbolically reinforcing its absence from Catherine's thoughts.

Characterization

- Nelly portrays **Catherine** as a thorn that only appears to have softened with marriage because she is so warmly embraced by the Lintons. However, when her desires are thwarted, Catherine manipulates Edgar and Isabella with "fits of passion" and by teasing—machinations that set off a chain of events that dooms the residents of both houses.
- **Edgar Linton** is portrayed as a calm, loving person, who deals with his wife's moods by humoring her and rejoicing when she is not depressed. His manners are no match for her fiery personality, however. "For the space of half a year the gunpowder lay as harmless as sand, because no fire came near to explode it," until Heathcliff, whose soul is made of fire, sets the lives of all involved ablaze because of his passion for Catherine.
- **Heathcliff** has been transformed in the three years of his absence. The contrast between him and Linton is stark: "He had grown a tall, athletic, well-formed man; beside whom, my master seemed quite slender and youth-like. . . . A half-civilized ferocity lurked yet in the depressed brows and eyes full of black fire."
- **Isabella** is eighteen and immature. Her naiveté makes her an instrument for Heathcliff's revenge. ***Ironically,*** marriage to Heathcliff toughens Isabella, who even fantasizes about killing him. Her escape from Wuthering Heights and her decision to live without her husband and raise their son alone shows courage that was particularly uncommon for that time period.
- **Hareton,** at this point in the novel, is also a tool of vengeance and a bargaining chip for Heathcliff. Nelly's visit to the Heights casts light on the effect that living in a loveless home has had on the boy. Nelly, who has bribed Hareton with an orange to extract information, describes how "[h]e hesitated, and then snatched it from my hold, as if he fancied I only intended to tempt and disappoint him."

Foreshadowing

- After Heathcliff's first visit to Thrushcross Grange, Nelly mused, "[H]e had better have remained away." When the visits increase she "felt that God had forsaken the stray sheep there to . . . an evil beast . . . waiting for his time to spring and destroy."
- Catherine warns the vulnerable Isabella that Heathcliff would crush her "like a sparrow's egg," and when Isabella finally escapes his tyranny she confesses to Nelly, "I gave him my heart, and he took and pinched it to death, and flung it back to me."
- In her delirium, Catherine speaks of being with Heathcliff on the moors as children daring ghosts to come to them. She promises not to rest till Heathcliff is with her, even if they bury her twelve feet deep. This foreshadows her death and Heathcliff's haunted life without her.
- When Edgar seeks custody of Hareton, Heathcliff refuses and threatens that if Edgar were to raise Hareton, he, Heathcliff, would find his own son Linton to replace Hareton. In the future, Heathcliff *will* acquire Linton and arrange for him to replace Hareton as master of Wuthering Heights.

Style
Gothic elements

- Heathcliff hangs Isabella's dog the night of their elopement. Though Nelly rescues the dog, this incident provides another example of Heathcliff's

Study Guide | 23

Plot Synopsis and Literary Elements (cont.)

cruelty; his treatment of the dog is symbolic of his treatment of people.
- When Catherine is in her room, seemingly losing her senses and raving to Nelly, she "sees" Wuthering Heights though it is not visible from Thrushcross Grange. She also "sees" Heathcliff and speaks to him about her obsessive longing for him, threatening to haunt him after she is dead and buried.
- With a touch of the Gothic supernatural, Cathy reiterates her longing for Heathcliff: "I only wish us never to be parted: and should a word of mine distress you hereafter, think I feel the same distress underground. . . ."
- With Gothic violence, Heathcliff describes what he would do to Edgar if he was sure Catherine felt nothing for him: "The moment her regard [for Edgar] ceased, I would have torn his heart out, and drunk his blood!"
- Heathcliff's anguish over Cathy's death is Gothic in tone: "Catherine Earnshaw, may you not rest as long as I am living! You said I killed you—haunt me, then! The murdered do haunt their murderers. I believe—I know that ghosts have wandered on earth. Be with me always—take any form—drive me mad!"

Theme
Love as a nurturing and creative force versus love as an all-consuming and destructive force
- Heathcliff's love affords him no generosity. Of Edgar he says, "If he loved with all the powers of his puny being, he couldn't love as much in eighty years as I could in a day . . . It is not in him to be loved like me."
- Catherine's death fuels Heathcliff's obsessive hatred of everyone who has brought him pain, giving him license to destroy Hindley and Isabella and plot against Edgar.
- In contrast, Edgar's response to the loss of his beloved is to focus his attention on their child and even to attempt to raise Heathcliff's child to protect him from his father's anger.

The destructive power of revenge
Hindley's downfall is the aftereffect of his own plan of revenge against Heathcliff.

Chapters XVIII–XXV

Plot Synopsis
The next twelve years are the happiest for Nelly. She is absorbed in bringing up Cathy, whose protective father does not allow her to travel beyond the park surrounding the Grange by herself. Cathy has grown curious about the outside world by the summer of 1797, when Edgar is summoned to London. Isabella is dying and needs to entrust her son Linton to him. In Edgar's absence, thirteen-year-old Cathy takes three of her dogs and secretly rides her pony into the hills.

Nelly finds her at the Heights, laughing and chattering with Hareton, a handsome but obviously uneducated boy. When Cathy mistakes Hareton for a servant and orders him to fetch her pony, he curses her. A servant reveals that Hareton is Cathy's cousin. Nelly convinces Cathy to say nothing to her father about the visit.

A letter announces Isabella's death, and Edgar returns from London soon after with Linton. Six months younger than Cathy, he is delicate and fretful. Shrinking from her enthusiastic welcome, he prefers to lie on the sofa while the others take tea. That evening Joseph arrives after both children are in bed. Heathcliff has sent Joseph to fetch his son. Knowing he cannot overcome Heathcliff's claim, Edgar reluctantly agrees to bring the boy the next day.

Edgar has Nelly deliver Linton in the early morning so Cathy does not know where he is. She is to be told

Plot Synopsis and Literary Elements (cont.)

Wuthering Heights

only that his father sent for him. Linton is bewildered and reluctant to leave, since Isabella never mentioned his father. When Heathcliff sees his sickly son, he is scornful, admitting in the presence of the young boy that he is "bitterly disappointed with the whey-faced whining wretch" and despises him because he looks so like his mother. Still, Heathcliff tells Nelly he intends to educate him and raise him as a gentleman and prospective owner of the Grange. As Nelly leaves, she hears Linton's frantic pleas not to be left behind.

Heathcliff's housekeeper advises Nelly of Linton's progress, and Nelly keeps Edgar informed. Heathcliff mostly avoids Linton and the child remains sickly and hard to please. Life at the Grange remains peaceful and pleasant until Cathy's sixteenth birthday. To amuse herself on this beautiful day, Cathy asks Nelly to walk with her on the moors. Cathy wanders ahead and when Nelly catches up, Cathy is being detained by Hareton and Heathcliff, as she is on their property. When Cathy asks Heathcliff if Hareton is his son, he invites her to his house to meet his real son, someone she already knows. Nelly rebukes Heathcliff; she knows this will lead to trouble. He acknowledges that he wants Cathy and Linton to fall in love and marry. He is being generous, he claims, since Linton, not Cathy, will inherit Thrushcross Grange.

Arriving at the Heights, Cathy greets Linton with kisses and addresses Heathcliff as "uncle." She wants to visit every day, but Heathcliff cautions secrecy, telling her that her father has quarreled with him. When Heathcliff tells Linton to show Cathy the garden and stables, the boy is reluctant to make the effort, and Heathcliff finally calls Hareton, who has been outside making himself presentable. Cathy and Hareton leave to tour the grounds. Heathcliff admits to Nelly that Hareton is a better person than his son but that he cannot love Hindley's son. Eventually, having grown resentful of being left behind, Linton finally joins Cathy and Hareton outside. When Hareton admits that he is unable to read the inscription above the door, both his companions, but especially Linton, taunt him about his ignorance and rough manners.

The next day Cathy tells her father about the visit. After he explains how badly Heathcliff treated both Isabella and Hindley, he orders her to never return or to contact anyone at the Heights. Cathy seems to accept these restrictions but later begs Nelly to help get a note of explanation to Linton. Nelly refuses, so Cathy secretly arranges for the milkboy to convey letters. Weeks later Nelly discovers Cathy's secret collection. She is shocked to find they are love letters and suspects the words come from someone other than Linton because the ideas reflect maturity Linton does not have. When confronted, Cathy confesses she loves Linton. She begs Nelly to burn the letters rather than tell her father. But Nelly does both, and the next day Edgar writes Heathcliff that no more notes should come to Cathy.

During the autumn harvest, Edgar catches a cold that affects his lungs and confines him to the house for the rest of the winter. Walking the grounds with Nelly, Cathy says she fears her father will die young like her aunt and she will be alone. Nelly assures Cathy her father will survive for a long time and advises Cathy to stay cheerful and never again mention Linton and Wuthering Heights to him. They hear a horseman approach. It is Heathcliff. He tells Cathy that Linton is dying of love for her and if she does not visit to restore his health he will send Edgar Cathy's love letters. Nelly hurries Cathy back to the Grange, but decides not to worry Edgar about the incident. Nelly agrees to go with her the next day to the Heights.

They travel through a freezing drizzle. At the Heights they find Joseph sitting by the fire, ignoring Linton's calls for more coal. Cathy rushes to kiss Linton but he turns away, claiming it takes his breath. Obviously feverish, he complains that having to write his long letters was tiring, scolding Cathy for not visiting more. Cathy assures him she loves him. Linton begs her to visit when Heathcliff is away. He needs her to take care of him and wants her to be his wife so she will love him more than her father. Linton is revealed to be both ill and very manipulative. Nelly says they will never

Plot Synopsis and Literary Elements (cont.)

Wuthering Heights

return, but Cathy merely smiles and tells Nelly she is not her jailer. Cathy is certain she can restore Linton to health and "make such a pet of him."

Nelly does not tell Edgar where they have been. The next day she is sick from the cold and rain and as a result is confined to the house for three weeks. Nelly eventually discovers Cathy returning from a secret visit to the Heights. Cathy admits that she goes almost every day.

She tells Nelly that on her first visit, she and Linton spent a pleasant time talking about their ideas of perfect happiness. The next day Hareton, eager to impress her, reads his name above the front door. But when he can read nothing else Cathy calls him a dunce and he leaves mortified. Nelly interrupts the story to reprimand Cathy's bad manners. Unconcerned, not realizing she was the cause of Hareton's behavior, she describes how Hareton interrupted her visit with Linton, grabbing him and cursing them both. Linton's attempt to fight back produced a fit of coughing and blood rushed from his mouth. Fearing Linton had died, Cathy returned several days later. Linton has recovered and blames Cathy for all the problems. Two days later, they reconcile after he apologizes and expresses love. Cathy begs Nelly not to tell her father what has happened, but she does. Alarmed, Edgar again forbids Cathy to go to the Heights. To comfort her, he writes Linton, inviting him to visit the Grange.

Nelly interrupts her story to observe she could not have predicted she would be describing the previous year's events to Lockwood, and asks if he has an interest in Cathy. Lockwood admits he may love her but is not willing to risk his tranquillity to pursue her. He returns Nelly to her narrative.

Nelly tells Edgar that the sickly Linton will probably die, and Edgar admits that he himself feels death approaching. He does not care that Heathcliff may inherit Thrushcross Grange; what he really wants is to see Linton become a worthy husband. Since he knows that will never happen, he would rather that Cathy die before he does than that his death place her in Heathcliff's power.

On Cathy's seventeenth birthday, Edgar writes Linton, asking him to visit the Grange. In a reply apparently dictated by Heathcliff, Linton speaks of his love for Cathy and asks Edgar to ride and meet them halfway on the moors. Edgar is too ill but suggests the possibility of meeting in summer and encourages Linton to keep writing. By summer, Cathy persuades Edgar to allow her to ride out under Nelly's guardianship to meet Linton. Edgar has set aside a portion of his income for his daughter, but he knows that marriage to Linton is the only way she can keep her home. He remains unaware, however, of his nephew's precarious health.

Nelly admits to Lockwood that at the time she did not suspect that Heathcliff was compelling his dying child to express love for Cathy as part of "his avaricious and unfeeling plans."

Literary Elements

Point of View: Nelly is the primary narrator of these chapters, but young Cathy Linton also provides first-person narration when she tells of her secret trip to Wuthering Heights. She proves honest in this task, admitting that often the visits with her cousin Linton were unpleasant.

Characterization

- **Linton's** character grows in the direction that it is first described as "sickly peevishness." He resembles his father "not a morsel," but he resembles Edgar and Isabella sufficiently to anger Heathcliff. The contempt with which Heathcliff regards his son shapes Linton's already unpleasant personality into one completely self-centered.

  While Cathy and Linton are courting, they discuss their ideas of heaven. Linton's ideal is lying all day on the moors while looking at a cloudless blue sky and listening to the larks. Cathy's heaven is swinging in a tree, with the wind blowing white clouds about the sky and a myriad of birds

Plot Synopsis and Literary Elements (cont.)

Wuthering Heights

singing along with the sounds of the woods and water. His static dream compared to her dynamic one highlights their different characters.

- **Cathy** is beautiful, with features of both the Earnshaws and the Lintons. She bears many similarities to her mother: she is high-spirited, intensely loyal, with a tendency to be saucy. Nelly makes a distinction between Cathy and her mother. "[H]er anger was never furious: her love never fierce. It was deep and tender." Cathy suffers from the conflict between her compassionate, sympathetic attraction to Linton, with whom she has little in common, and her immediate attraction to the uneducated Hareton, who shares some of her interests (outdoor pursuits, nature).
- **Hareton** "is gold put to the use of paving-stones," according to Heathcliff, and Heathcliff says his son Linton "is tin polished to ape a service of silver." Hareton's rough demeanor, the result of Heathcliff's depriving him of education and opportunity, is reminiscent of Heathcliff in his youth. **Ironically,** Hareton, who has every reason to hate him, considers Heathcliff "the one friend he has in the world."
- **Heathcliff's** manipulations seem to know few bounds. Positive aspects of his character, however, are shown in his interaction with Nelly and his sympathy for Hareton. Nelly is his confidant; only through her does the reader gather that Heathcliff feels pity for Hareton—though he is responsible for his degradation. Early critics called Heathcliff inhuman, but these contradictions in character help to humanize him.

Symbolism: Raised within the confines of the Grange, young Cathy by age thirteen becomes curious about what "lies on the other side" of her world. "The abrupt descent of Penistone Craggs particularly attracted her notice; especially when the setting sun shone on it and the topmost heights, and the whole extent of landscape besides lay in shadow." Her attraction to this *heath cliff* will inspire her adventure beyond the grounds and thus her first meeting with Heathcliff and her cousin Hareton.

Catherine gave birth and died at the vernal equinox, symbolically a point when dark and light, death and life, are in equal balance. On the anniversary of this date, Cathy's sixteenth birthday, she meets Heathcliff and Linton again, setting in motion Heathcliff's plan for revenge against Edgar.

Foreshadowing: As Nelly and Cathy wander about the moors on Cathy's sixteenth birthday, the young girl is anxious to move beyond the perimeters set for her. Nelly tells Lockwood, "It's a pity she could not be content." Cathy's enthusiasm takes her to Heathcliff's Wuthering Heights.

Several clues to the natural attraction between Hareton and Cathy hint at a forthcoming friendship:

- In Chapter XVIII, Hareton rescues Cathy's dogs from a dogfight; "open[s] the mysteries of the Fairy cave and twenty other queer places"; and invites her into Wuthering Heights where he then entertains her. Even when she offends him by supposing he is a servant, it is Hareton who makes a peace offering.
- On her second visit three years later, despite her reintroduction to her cousin Linton, Cathy whispers something to Heathcliff that causes him to declare that it appears Hareton shall be her favorite of them all.
- Hareton learns to read to impress Cathy.

Plot Synopsis and Literary Elements (cont.)

Chapters XXVI–XXXIV

Plot Synopsis

On their first summer ride to join Linton, Nelly and Cathy find him a quarter mile from his front door without his horse and lying on the ground. He can barely walk. He is apathetic and listless, but rouses when Cathy tires of the one-way conversation and declares they must go. He begs her not to reveal his declining health to her father, because this revelation would provoke more harsh treatment by Heathcliff. After the exhausted boy dozes, Cathy tells Nelly that Heathcliff has likely forced this meeting. Despite her disappointment in the visit, Cathy agrees to return in a week. Edgar is not told about Linton's rapid decline.

During the next week, Edgar's health deteriorates rapidly and Cathy realizes he will die soon. Yet at his insistence, Nelly and Cathy set out to meet Linton, as promised. He acts strangely and after Cathy tells him they cannot stay long because of her father's condition, Linton grabs her skirt and begs her to stay. After Heathcliff appears, Nelly suspects Linton's behavior is caused by fear. Heathcliff has heard Edgar is near death, which Nelly confirms. Heathcliff says he fears Linton will die first, then berates his son for groveling. He asks Cathy and Nelly to help Linton inside the house but they decline, explaining that Edgar has forbidden it. An irate Heathcliff begins to escort his son, Linton clings to Cathy, and they all go inside. Heathcliff locks them in. Unafraid, Cathy demands the key. Heathcliff is momentarily startled by her boldness, a reminder of her mother. Cathy snatches the key but he eventually recovers it, then repeatedly slaps her on both sides of the head. Heathcliff declares he will be her father tomorrow and will beat her again if her temper flares.

When Heathcliff leaves, Linton explains to Cathy that Heathcliff wants them to marry the next morning, then return to the Grange as a couple. Cathy's response is to look for an escape route. Linton begs her to stay and save him by doing as his father pleases. He is crying when Heathcliff returns and locks him in his room.

Cathy promises to marry Linton if Heathcliff will let her return immediately to her father, but her pleas are to no avail. Heathcliff locks Cathy and Nelly in Zillah's bedroom. The next morning Heathcliff allows Cathy to leave the room but not the Heights. Linton and Cathy are married. For four days Nelly remains inside, seeing only Hareton, who brings food but no sympathy.

On the fifth day, Zillah returns. She has heard rumors that Nelly and Cathy had drowned in the marsh. Nelly is to go back to the Grange and Heathcliff promises to send Cathy in time for her father's funeral. When Nelly asks if Edgar is dead, Zillah says he may last one more day. Nelly finds Linton alone downstairs, contentedly sucking sugar candy. He tells her Cathy is upstairs but cannot leave. His father has told him Cathy is interested only in his money and that he should be harsh to her now that she is his wife. When Nelly reminds Linton of Cathy's kindness, he complains that her crying keeps him awake. Disdainfully, he describes Cathy's attempts to bribe him, offering everything dear to her, including a gold locket containing both parents' pictures. He is annoyed that she now refuses to speak to him. He refuses to take Nelly to Cathy. Nelly leaves, having resolved to return with a rescue party.

Upon reaching the Grange, Nelly can see Edgar's drastic decline. After she explains what happened, Edgar orders her to send for the lawyer. He wants to put Cathy's money in trust for her use and that of her children. Nelly sends for Mr. Green, the lawyer, and also dispatches four men to retrieve Cathy. Mr. Green sends word that he will come before morning; the men return with word that Cathy is too ill to travel. At three in the morning, Nelly hears a knock at the door. It is Cathy. Nelly implores her not to upset Edgar but to say she expects to be happy as Linton's

Plot Synopsis and Literary Elements (cont.)

Wuthering Heights

wife. Cathy hides her despair from her father and he dies blissfully in her arms. When the lawyer arrives, Nelly realizes the delay was at Heathcliff's command. The lawyer orders all servants except Nelly dismissed and tries to have Edgar buried in the Linton plot instead of next to Catherine. Here, Nelly successfully thwarts Heathcliff by pointing out that Edgar's will specifies his burial place and cannot be overturned.

The evening after Edgar's funeral, Cathy and Nelly talk about Cathy's future. Cathy wants to move with Linton to the Grange but suspects Heathcliff will not allow it. Heathcliff enters unannounced and confirms that he has come to take Cathy back to Wuthering Heights, where he hopes she will be a dutiful daughter and look after Linton. When Nelly asks why the couple cannot remain at the Grange, Heathcliff says he is seeking a tenant and, besides, Cathy should earn her keep.

Before leaving to gather her things, Cathy states that she loves Linton and that Heathcliff cannot make her hate him. Heathcliff responds that it is Linton's nature to hate because he has inherited his father's nature. Cathy says she is glad that her nature is better so she can forgive Linton's faults. She reminds Heathcliff he has no one to love him.

After Cathy leaves, Heathcliff tells Nelly he wants the portrait of Catherine sent to the Heights. He confides that he visited Catherine's open grave before preparations began for Edgar's burial. He persuaded the sexton to open her coffin and saw that her face was unchanged. He loosened the side of her coffin farthest from Edgar's grave and bribed the sexton to pull the side away entirely when he is buried beside her. His own coffin will have a removable panel on the side next to Catherine's, so that he can be "united" with her in death.

Nelly chides Heathcliff for disturbing the dead. He tells her that Catherine has disturbed him night and day for eighteen years. He explains that the night after Catherine's funeral he dug at her grave, determined to have her in his arms again. As the spade was about to crack the coffin lid, he heard a sigh above him and felt Catherine's presence.

Heathcliff forbids Nelly to visit Cathy at the Heights. Nelly meets on the moors with Zillah and learns that Heathcliff has ordered that Cathy be left to fend for herself and that she has stayed in Linton's room to care for him. Cathy asked for a doctor for Linton, but Heathcliff refused, saying the boy's life is worth nothing and no one should assist her. Linton died. Only Hareton seemed moved, not as much for Linton as for Cathy. She remained in her room for two weeks, repelling all attempts at friendliness. Heathcliff visited once to show Linton's will to Cathy. It left everything, including Cathy's personal fortune, to Heathcliff.

She eventually rejoined the others in the household when Heathcliff was away. Hareton had made himself presentable, but Cathy showed him nothing but scorn. When she discovered some books on a shelf too high to reach, Hareton got them for her and then looked over her shoulder as she read. She refused his request to read aloud, claiming it was hypocritical since no one had shown her any kindness while Linton was dying. Hareton claimed he had offered to help but was overruled by Heathcliff. Cathy even grew snappish with Heathcliff and more venomous when he beat her.

After hearing this account, Lockwood plans to meet with Heathcliff in the next few days. He does not intend to remain at Thrushcross Grange for the remainder of his lease and needs to tell Heathcliff to look for a new tenant.

Bearing a note from Nelly for Cathy, Lockwood is shown inside by Hareton, who is now extremely handsome. When told Heathcliff will not be home until mealtime, Lockwood resolves to wait. Cathy appears less spirited but still beautiful. Lockwood tries to drop Nelly's letter in her lap undetected, but when she pushes it to the floor he is forced to say it is for her from Nelly. Hareton snatches it, saying Heathcliff should see it first, but relents and gives it to Cathy when she

Plot Synopsis and Literary Elements (cont.)

Wuthering Heights

begins to cry. After reading it, she tells Lockwood she cannot respond because she has no materials with which to write. She accuses Hareton of stealing her books, because she has seen them in his room. Hareton denies this. Lockwood defends him, telling Cathy her cousin is trying to learn to read to emulate her achievements. She mimics his blunders. Hareton can barely hide his mortification. He quickly fetches her books. She again mimics his poor reading, provoking him to throw the books into the fire. Lockwood quietly sympathizes with Hareton's frustration.

As Hareton storms out, Heathcliff enters. When Lockwood tells Heathcliff his intention to return early to London, Heathcliff responds that he will still expect full payment from Lockwood even if he is not in residence at the Grange. Irritated, Lockwood offers to pay the balance on the spot, but Heathcliff will not accept it. After banishing Cathy, Heathcliff invites his tenant to stay for dinner. Lockwood finds it cheerless and leaves soon after. Thwarted in his attempt at a last glimpse of Cathy, Lockwood rides back, daydreaming of life with Cathy in London.

It is September 1802, nine months later. While hunting on a friend's estate, Lockwood impulsively decides to visit the Grange. Expecting to find Nelly, he is greeted by an old woman who says Nelly has moved to the Heights. Flustered, he sets out and is surprised to find the gate at Wuthering Heights unlocked and the fragrance of flowers in the air. Through an open window he sees Cathy teaching Hareton to read and bestowing kisses on him. They are a well-dressed, handsome couple, and he regrets throwing away the chance to woo Cathy. When they leave to walk on the moors, he enters the house, where Nelly is sewing. She is delighted to see Lockwood. He learns that Heathcliff is dead and is eager to learn the details. Nelly readily complies, beginning with her arrival at Wuthering Heights.

Heathcliff requested Nelly's presence to keep Cathy out of his sight as much as possible. At first Cathy is pleased but soon grows irritable and lonely. Her chief amusement is quarreling with Joseph and confronting Hareton about his stupidity and idleness. Eventually she regrets having tormented him and tries to make friends, but he avoids her. She even leaves books for him, but Hareton avoids the bait.

While Hareton is recuperating from a shooting accident, Cathy tells him she would now be glad to acknowledge him as her cousin. Hareton at first repels her. Cathy wraps a handsome book and tells Nelly to present it with the message that if he accepts it she will teach him to read properly and even if he refuses she will never tease him again. After some hesitation, Hareton unwraps the book and they become friends at last.

As Cathy teaches Hareton to read, Nelly notices a new spirit and nobility in him. When Heathcliff discovers Cathy and Hareton reading together, he is disarmed by their resemblance to Catherine. Cathy has her eyes; but Hareton's similarity to his aunt is striking. Heathcliff takes the book and motions them to leave. He detains Nelly and tells her that he has lost the will for further revenge. He feels a "strange change" approaching. He confesses he sees evidence of Catherine everywhere. Nelly now believes what Joseph has said, that Heathcliff's conscience has turned his heart "to an earthly hell."

For several days Heathcliff refuses to join the others at meals and eats only once a day. One night Nelly hears him leave the house. In the morning he is still gone. After breakfast, Cathy encounters Heathcliff and tells Nelly he appeared surprisingly cheerful. Nelly detects "a strange joyful glitter in his eyes" and asks him if he has received good news. He tells her he is within sight of heaven and asks her to stop prying. He eats nothing that day. That evening when Nelly brings him a candle and supper, she discovers the fire almost out and all the windows open. Heathcliff is smiling and pale. He refuses to eat the next day and sleeps in Catherine's bed that night. Nelly tries to coax him to eat, but he eventually becomes irritated and leaves the house, not returning until after midnight. Nelly hears him pacing and muttering Catherine's name as if she were present. He tells Nelly

Plot Synopsis and Literary Elements *(cont.)*

Wuthering Heights

he intends to send for Mr. Green at daybreak to order a will and asks her to remind the sexton to obey his orders about his funeral: He wants no minister there. Nelly hears him groaning through the night and into the morning. Concerned, she sends Hareton for the doctor, but Heathcliff refuses to let him in. The next night it rains heavily. In the morning, she discovers Heathcliff, his eyes wide open, simultaneously smiling and sneering. He is dead. Only Hareton sincerely grieves. The funeral takes place as Heathcliff wishes, scandalizing the whole neighborhood.

Nelly tells Lockwood she has heard reports that villagers have seen Heathcliff's ghost near the church and on the moor. She has also met a shepherd boy who claimed to have seen the ghosts of Catherine and Heathcliff walking near Wuthering Heights. She will be glad when Hareton and Cathy move to the Grange after their marriage on New Year's Day. When Lockwood asks what will happen to Wuthering Heights, Nelly says Joseph will remain in the kitchen quarters and the rest of the house will be closed.

Lockwood decides to leave as secretly as he entered in order not to disturb the young people. He gives Nelly a remembrance and tosses a sovereign at Joseph's feet. On his way back to the Grange, he stops at the churchyard to look at the three graves alongside the moor. Catherine's headstone is half buried "in the heath"; Edgar's has moss at its foot; Heathcliff's is still bare. Lockwood wonders "how any one could ever imagine unquiet slumbers for the sleepers in that quiet earth."

Literary Elements

Point of View: Nelly's narration is augmented by that of Zillah, and at the novel's conclusion Lockwood resumes his role as narrator.

Structure: The symmetry of *Wuthering Heights* becomes apparent in this final section. By the end of Chapter XXX, the story has come full circle: the reader has reached the point in time at which the novel began. Lockwood's visits to Wuthering Heights, coupled with his narration, stand as prologue and epilogue to Nelly Dean's story. Brontë, as she did at the novel's beginning, alternates between Lockwood and Nelly and between present tense and flashback.

Characterization
- **Nelly's** character is developed through **indirect characterization.** Significant details about this ever-present character are evident in her rapport with Heathcliff. He confides in her, and his behavior to her is civil, in contrast to his interaction with others. Heathcliff's confidence in Nelly lends credibility to her role as narrator.
- **Heathcliff** in these chapters reaches the height of cruelty and then ebbs into a more pathetic character. Driven to revenge, he has the audacity to lock Cathy and Nelly in his house to force a marriage between Cathy and Linton; after Edgar's death he brings Cathy back to the house to be nothing more than a servant; he punishes his son by silence and intimidation for his part in helping Cathy escape to Thrushcross Grange; and then he does little to prevent Linton's death. Ultimately however, his plans are undermined by the memory of Catherine that Cathy and Hareton evoke. When he sees Catherine's corpse on the night of Edgar's burial and her likeness in the very persons he has tried to destroy, he loses the will for revenge. With this realization, Heathcliff begins to pursue his second great goal: union with Catherine.

  Interestingly, Heathcliff is not alone in his desire to be reunited with Catherine. **Edgar's** final words to his daughter are "I am going to her; and you darling child shall come to us." The dignity of his death again shows Edgar as a **foil** to Heathcliff. The undying passion that Heathcliff felt for Catherine was his excuse for destroying lives; that passion was Edgar's reason for nurturing Cathy's life.
- **Cathy,** bitter and lonely, whom Lockwood encountered in Chapter II is the product of the circumstances surrounding the deaths of her father and Linton. Heathcliff's neglect of the dying Linton and

Plot Synopsis and Literary Elements (cont.)

Wuthering Heights

then the grieving Cathy is almost enough to harden her heart. A growing affection for Hareton and her spirit of forgiveness reverse these effects. Cathy, unlike the other inhabitants of the Heights, openly defies Heathcliff; she goes so far as to threaten Heathcliff with his own creation—Hareton—having won him over with a gentle love.

- **Hareton,** though in love with Cathy, would have done no harm to Heathcliff: he is loyal to him, which is a part of his nobility that a forced ignorance has not effaced. The reader learns along with Cathy that Hareton often spoke on her behalf to Heathcliff. Hareton's frustration at learning to read and trying to impress Cathy is reminiscent of Heathcliff's attempts to improve his appearance and manners for Catherine's sake. Heathcliff explains to Nelly, "Hareton seemed a personification of my youth, not a human being: I felt to him in such a variety of ways that it would have been impossible to have accosted him rationally. . . . Hareton's aspect was the ghost of my immortal love, of my wild endeavors to hold my right, my degradation, my pride, my happiness, and my anguish—" The sympathy Heathcliff feels for Hareton and the vision of Catherine he sees in Hareton will unravel his plans for revenge.
- **Linton** is the most **static** of the novel's characters. His one act of defiance and selflessness did afford Cathy the opportunity to escape Wuthering Heights to be present at her father's death. Linton's treatment of Cathy after their marriage, though, is another manifestation of a personality warped by his father's scorn.

Symbolism
- Cathy escapes through her mother's old bedroom window, showing the same sort of high spirit her mother had.
- Details of setting, such as the sinking sun and rising moon and the open window, reinforce the mystical nature of Heathcliff's "strange change."
- Like Catherine before him, Heathcliff experiences a mental transformation that brings about an end to the physical life.
- It is nine months between Lockwood's departure and return, and a new liveliness has been born at Wuthering Heights during that time.
- This new liveliness is represented in the flowers and fruit trees planted before the house, and open doors and windows.
- The wedding for Hareton and Cathy is set for New Year's Day.

Style
Gothic elements
- Heathcliff says, "Had I been born where laws are less strict and tastes less dainty, I should treat myself to a slow vivisection of those two [Cathy and Linton], as an evening's amusement."
- The night of Edgar's burial, Heathcliff asks the sexton to remove the coffin lid so he can see Catherine's face. The fact that it has not deteriorated is **symbolic** of her state of grace (at least in Heathcliff's mind)—it is a traditional belief that saints' bodies remain uncorrupted.
- Heathcliff has the sexton loosen the side of her coffin so that in death their dust may mingle.
- Heathcliff is convinced that Catherine's spirit was with him the night of her burial, when he planned to open her coffin and hold her in his arms one final time. In a prime example of the Gothic supernatural, her presence came to him, making this unnecessary.
- "Is he a ghoul, or a vampire?" muses Nelly at Heathcliff's nighttime ramblings.
- Upon Heathcliff's death, Joseph exclaims that "[t]he devil's carried off his soul." By the time Lockwood returns to the Heights, it is local lore that Heathcliff and Catherine wander the moors as ghosts.

Plot Synopsis and Literary Elements (cont.)

Theme

Love as a nurturing and creative force versus love as an all-consuming and destructive force

- Cathy, in a defiance that will eventually save her, informs her father-in-law, "Linton is all I have to love in the world, and, though you have done what you could to make him hateful to me, and me to him, you cannot make us hate each other!"
- It is Cathy's nurturing love of Hareton that transforms both the sullen young man and Wuthering Heights.
- At his death, Edgar looks forward to being reunited with Catherine, revealing no bitterness for the love he must have known she felt for Heathcliff.
- Heathcliff destroys his body, allowing it to wither away so that he can be reunited with the spirit of Catherine.

The destructive power of revenge

- The warped nature of Linton's personality is an effect of Heathcliff's revenge.
- Heathcliff becomes a fiend, one Joseph even supposes is possessed by the devil, in his vengeance. He kidnaps, mentally and physically abuses the inhabitants of Wuthering Heights, and allows his son to die, caring little when he does.

Passion over reason

- Twice Heathcliff is blinded by rage and ready to strike Cathy, only to be halted by her resemblance to her mother.
- Ultimately, it is Heathcliff's passion that drives him to starvation in order to be reunited with his love, Catherine.

Reader's Log: Model

Reading activity In your reader's log you record your ideas, questions, comments, interpretations, guesses, predictions, reflections, challenges—any responses you have to the books you are reading.

Keep your reader's log with you while you are reading. You can stop at any time to write. You may want to pause several times during your reading time to capture your thoughts while they are fresh in your mind, or you may want to read without interruption and write when you come to a stopping point such as the end of a chapter or the end of the book.

Each entry you make in your reader's log should include the date, the title of the book you are reading, and the pages you have read since your last entry (pages ___ to ___).

Example

Sept. 21

<u>Fahrenheit 451</u>
pages 3 to 68

This book reminds me a lot of another book we read in class last year, <u>1984</u> by George Orwell. They're both books about the future—<u>1984</u> was written in the 1940s so it was the future then—a bad future where the government is very repressive and you can be arrested for what you think, say, or read. They're also both about a man and a woman who try to go against the system together. <u>Fahrenheit 451</u> is supposed to be about book censorship, but I don't think it's just about that—I think it's also about people losing their brain power by watching TV all the time and not thinking for themselves. <u>1984</u> did not have a very happy ending, and I have a feeling this book isn't going to either.

Writing for an audience Exchange reader's logs with a classmate and respond in writing to each other's most recent entries. (Your entries can be about the same book or different ones.) You might ask a question, make a comment, give your own opinion, recommend another book—in other words, discuss anything that's relevant to what you are reading.

Or: Ask your teacher, a family member, or a friend to read your most recent entries and write a reply to you in your reader's log.

Or: With your teacher's guidance, find an online pen pal in another town, state, or country and have a continuing book dialogue by e-mail.

Reader's Log: Starters

When I started reading this book, I thought . . .

I changed my mind about . . . because . . .

My favorite part of the book was . . .

My favorite character was . . . because . . .

I was surprised when . . .

I predict that . . .

I liked the way the writer . . .

I didn't like . . . because . . .

This book reminded me of . . .

I would (wouldn't) recommend this book to a friend because . . .

This book made me feel . . .

This book made me think . . .

This book made me realize . . .

While I was reading I pictured . . . (Draw or write your response.)

The most important thing about this book is . . .

If I were (name of character), I would (wouldn't) have . . .

What happened in this book was very realistic (unrealistic) because . . .

My least favorite character was . . . because . . .

I admire (name of character) for . . .

One thing I've noticed about the author's style is . . .

If I could be any character in this book, I would be . . . because . . .

I agree (disagree) with the writer about . . .

I think the title is a good (strange/misleading) choice because . . .

A better title for this book would be . . . because . . .

In my opinion, the most important word (sentence/paragraph) in this book is . . . because . . .

(Name of character) reminds me of myself because . . .

(Name of character) reminds me of somebody I know because . . .

If I could talk to (name of character), I would say . . .

When I finished this book, I still wondered . . .

This book was similar to (different from) other books I've read because it . . .

This book was similar to (different from) other books by this writer because it . . .

I think the main thing the writer was trying to say was . . .

This book was better (worse) than the movie version because . . .

(Event in book) reminded me of (something that happened to me) when . . .

Study Guide | 35

Double-Entry Journal: Models

Responding to the text Draw a line down the middle of a page in your reader's log. On the left side, copy a meaningful passage from the book you're reading—perhaps a bit of dialogue, a description, or a character's thought. (Be sure to note the number of the page you copied it from—you or somebody else may want to find it later.) On the right side, write your response to the quotation. Why did you choose it? Did it puzzle you? confuse you? strike a chord? What does it mean to you?

Example

Quotation	Response
"It is a truth universally acknowledged, that a single man in possession of a good fortune must be in want of a wife." (page 1)	This is the first sentence of the book. When I first read it I thought the writer was serious—it seemed like something people might have believed when it was written. Soon I realized she was making fun of that attitude. I saw the movie Pride and Prejudice, but it didn't have a lot of funny parts, so I didn't expect the book to be funny at all. It is though, but not in an obvious way.

Creating a dialogue journal Draw a line down the middle of a page in your reader's log. On the left side, comment on the book you're reading—the plot so far, your opinion of the characters, or specifics about the style in which the book is written. On the right side of the page, your teacher or a classmate will provide a response to your comments. Together you create an ongoing dialogue about the novel as you are reading it.

Example

Your Comment	Response
The Bennet girls really seem incredibly silly. They seem to care only about getting married to someone rich or going to balls. That is all their parents discuss, too. The one who isn't like that, Mary, isn't realistic either, though. And why doesn't anyone work?!	I wasn't really bothered by their discussion of marriage and balls. I expected it because I saw the movie Emma, and it was like this, too. What I don't understand is why the parents call each other "Mr." and "Mrs."—everything is so formal. I don't think women of that class were supposed to work back then. And people never really work on TV shows or in the movies or in other books, do they?

Wuthering Heights

Name _____ Date _____

Group Discussion Log

Group members

....

....

....

....

....

Book discussed

Title: ....

Author: ....

Pages ____ to ____

Three interesting things said by members of the group

....

....

....

....

....

....

What we did well today as a group

....

....

....

....

What we could improve

....

....

....

Our next discussion will be on _____. We will discuss pages _____ to _____.

Glossary and Vocabulary

Wuthering Heights

- **Vocabulary Words** are preceded by an asterisk (*) and appear in the Vocabulary Worksheets.
- Words are listed in their order of appearance.
- The definition and the part of speech are based on the way the word is used in the chapter. For other uses of the word, check a dictionary.

Chapter I

misanthropist *n.:* person who hates or distrusts all other people; misanthrope

*** manifested** *v.:* revealed

flags *n.:* flat stones used for paving walkways

soliloquised *v.:* said to oneself

*** peevish** *adj.:* irritable; impatient

ejaculation *n.:* sudden forceful exclamation

surly *adj.:* bad-tempered; hostile and rude

penetralium *n.:* the innermost part, as of a temple, usually kept hidden

sundry *adj.:* various; miscellaneous

*** stalwart** *adj.:* strong and well built; sturdy

gypsy *n.:* member of a wandering people (typically with dark skin and black hair), found throughout the world and believed to have originated in India, often in history treated as outcasts and objects of discrimination

*** slovenly** *adj.:* untidy; careless in appearance

*** morose** *adj.:* gloomy; ill-tempered

impertinence *n.:* lack of proper manners or respect; inappropriateness

vis-à-vis *adj.:* (French) face to face with

*** tacit** *adj.:* silent; unspoken

physiognomy *n.:* facial features and expression

vexatious *adj.:* annoying; troublesome

loath *adj.:* unwilling; reluctant

prudential *adj.:* resulting from sound judgment, especially in practical matters

laconic *adj.:* using few words; brief

Chapter II

essay *adv.:* to attempt

in a pet *adv.:* in a state of ill humor

corrugated *adj.:* wrinkled; furrowed

assiduity *n.:* careful, constant attention

*** sagacity** *n.:* sound judgment; keen perception

reprobate *n.:* a corrupt person; one who is rejected by God as beyond salvation

*** malignity** *n.:* desire to do harm to others; intense ill will

ensconcing *adv.:* settling comfortably or snugly

postern *n.:* back door or gate; private entrance

copestone *n.:* top stone of a wall; used here to mean "final stroke" or "last straw"

miscreants *n.:* villains

virulency *n.:* spitefulness; malice

King Lear *n.:* title character of a tragedy by Shakespeare. In the first scene, Lear angrily threatens his loyal adviser, Kent.

Chapter III

*** vapid** *adj.:* dull; boring

dilapidation *n.:* shabbiness; state of ruin, as through neglect

garret *n.:* attic; rooms within an attic

asseverated *v.:* stated positively; asserted

owd Nick *n.:* Old Nick; Satan

lachrymose *adj.:* tearful

Seventy Times Seven: allusion to Matthew 18:22. When Peter wonders how many times a sinner should be forgiven, he suggests seven times, but Jesus disagrees, saying "until seventy times seven." Lockwood's nightmare stems from a literal interpretation of the biblical passage: Presumably, a person who keeps count need not forgive the 491st sin. Brontë, through Lockwood's dream, is satirizing people like Joseph who know the Bible thoroughly but fail to practice its message of forgiveness and love.

unhasp *v.:* unfasten

*** tenacious** *adj.:* holding firmly; clinging

the yell was not ideal: that is, was not merely an "idea"; Lockwood did not imagine it

Glossary and Vocabulary (cont.)

Wuthering Heights

maxillary *adj.:* having to do with the jawbone

changeling *n.:* child secretly exchanged for another, as by fairies. Such a child supposedly possessed supernatural qualities.

* **querulous** *adj.:* complaining; irritable

Grimalkin *n.:* an old, female cat; an allusion to Act I, Scene 1 of Shakespeare's *Macbeth*.

orisons *n.:* prayers

sotto voce *adv.:* (Italian) in an undertone so as not to be overheard

egress *n.:* the act of going out

* **impalpable** *adj.:* that cannot be felt by touching

Chapter IV

exotic *n.:* foreigner

indigenae *n.:* natives; those born in a region

dunnock *n.:* hedge sparrow. When Nelly says, "It's a cuckoo's sir, . . . Hareton has been cast out like an unfledged dunnock!" she is referring to the cuckoo's habit of laying its eggs in the nests of other birds. When the cuckoo chick hatches, it often pushes the mother bird's own chicks out of the nest.

Liverpool *n.:* a port in northwest England. In the years covered by the story, roughly 1771–1801, thousands of poor and homeless people lived in Liverpool; many were foreigners.

bairns *n.:* (Scottish) children

* **interloper** *n.:* intruder

Chapter V

who made the living answer by: who made a sufficient living by

Chapter VI

delf-case: case for storing and displaying china such as delftware, noted for its blue and white glaze. Delftware originated in the city of Delft in the Netherlands.

evincing *adj.:* indicating; showing plainly

cant *v.:* whine

vociferated *v.:* shouted loudly

execrations *n.:* curses

* **culpable** *adj.:* blameworthy

Lascar *n.:* sailor of India or East India

* **expostulating** *v.:* arguing or objecting to another's actions or views

negus *n.:* beverage of hot water, wine, lemon juice, sweetener, and spices

Chapter VII

beaver *n.:* heavy woolen cloth similar to felt

discomfiture *n.:* uneasiness; embarrassment

cant *adj.:* (dialect) bold; hearty

* **dour** *adv.:* in a sullen or gloomy manner

* **equanimity** *n.:* composure; evenness of temper

prognosticate *v.:* predict

Chapter VIII

consumption *n.:* disease that causes a wasting away of the body, specifically tuberculosis of the lungs

* **doggedly** *adv.:* stubbornly

dissipation *n.:* indulgence in pleasure to the point of harming oneself

* **hector** *v.:* bully

* **antipathy** *n.:* strong dislike

* **petulantly** *adv.:* irritably

* **imperiously** *adv.:* in an overbearing manner; arrogantly

assiduously *adv.:* diligently; carefully

waxed *v.:* became; grew

livid *adj.:* lead-colored; pale

* **consternation** *n.:* great fear or shock that makes one feel bewildered or helpless

timidity *n.:* shyness; lack of self-confidence

Chapter IX

vagaries *n.:* odd or unexpected actions

blasphemer *n.:* one who speaks disrespectfully of God or sacred matters

imprecations *n.:* curses

settle *n.:* long wooden bench with a high back

grat *v.:* wept

mither *n.:* (Scottish) mother

mools *n.:* mounds of earth over a grave

* **winsome:** attractive in a sweet way; charming

catechism *n.:* formal series of questions

injudicious *adj.:* showing poor judgment

Milo *n.:* Greek Olympic champion who lived around 520 B.C. The "fate of Milo" refers to being devoured by wolves.

Glossary and Vocabulary (cont.)

Wuthering Heights

bled *v.*: drew blood. To relieve fever, doctors formerly bled their patients, either by opening a vein or by applying leeches to the body.

* munificent *adj.*: very generous; lavish

Chapter X

dilatory *adj.*: slow; inclined to delay

blisters *n.*: substances used to create blisters on the skin, formerly used by doctors

phalanx *n.*: massed group of individuals, such as soldiers in close ranks

sizar's place *n.*: entry to Cambridge University on scholarship

averred *v.*: declared (something) to be true

sough *n.*: murmuring sound made by water; a boggy or swampy place; a small pond

beck *n.*: small stream

fastidiousness *n.*: tendency to be refined in a too dainty or oversensitive way, so as to be easily disgusted

presentiment *n.*: feeling that something unfortunate is about to occur; foreboding

covetousness *n.*: greed

abjured *v.*: gave up; renounced

furze *n.*: prickly evergreen shrub

* avarice *n.*: greed for riches

malevolence *n.*: ill will; spitefulness; malice

mitigating *v.*: making milder; moderating

mawkish *adj.*: sentimental in a weak, dull way

Chapter XI

propitiate *v.*: win or regain the good will of; appease; pacify

impudence *n.*: shameful boldness; disrespect

approbation *n.*: approval

intractable *adj.*: hard to manage; unruly; stubborn

ignominious *adj.*: humiliating; degrading

derision *n.*: contempt; ridicule

leveret *n.*: a young hare

stolidity *n.*: lack of emotion or sensitivity; impassivity

stoical *adj.*: calm and unflinching under suffering or distress

compunction *n.*: sharp feeling of uneasiness caused by a sense of guilt; remorse

peremptorily *adv.*: in an overbearing, arrogant manner

Chapter XII

pertinaciously *adv.*: stubbornly

pigeons' feathers *n.*: in folk belief, pigeon feathers supposedly prevented the soul from freeing itself of a dying person

elf-bolts *n.*: prehistoric arrowheads made of stone, believed in folklore to be used by malicious elves to kill cattle

paroxysm *n.*: sudden outburst; fit; spasm

kirk *n.*: church

Chapter XIII

* sanguine *adj.*: confident; optimistic

obviate *v.*: do away with; prevent

thible *n.*: a stick used for stirring

Chapter XIV

slattern *n.*: woman careless and sloppy in habits and appearance

importuned *v.*: urged repeatedly; (arch.) asked for urgently

alacrity *n.*: eager willingness

perspicacity *n.*: keen judgment; shrewdness

brach *n.*: (arch.) female hound

abject *adj.*: miserable; wretched; of the lowest degree

iteration *n.*: repetition

Chapter XVI

securing his estate to his own daughter, instead of his son's: Mr. Linton's will stipulated that his estate would pass from Edgar to Isabella in the event that Edgar died without a male heir.

heterodox *adj.*: departing from usual beliefs; unorthodox

Chapter XVII

* allayed *adj.*: relieved; lessened

incarnate *adj.*: being a living example of; made flesh

* recommenced *v.*: began again

* usurped *v.*: took or assumed, esp. by force or without right

preter-human *adj.*: more than one would expect of a person; superhuman

recapitulation *n.*: brief repetition; summary

basilisk *adj.*: spellbinding; allusion to a mythical lizardlike monster whose glance was supposedly fatal

* presumptuous *adj.*: arrogant; taking too much for granted

purgatory *n.*: a state or place of temporary punishment

Glossary and Vocabulary (cont.)

Wuthering Heights

* **inveterate** *adj.*: of long standing; established over a long period

Chapter XVIII

larch *n.*: kind of pine tree that loses its needlelike leaves annually

wisht *v.*: hush; shush

* **innuendoes** *n.*: indirect remarks, usually implying something derogatory; insinuations

comminations *n.*: threats or denunciations

near *adj.*: stingy; tightfisted

Chapter XIX

* **incipient** *adj.*: just beginning

* **trepidation** *n.*: anxiety; fearful uncertainty

Chapter XX

brown study *n.*: condition of being deeply absorbed in one's own thoughts, especially somber thoughts

cogitations *n.*: meditations; serious thoughts

ague *n.*: fit of shivering; fever

puling *adj.*: whimpering; whining

Chapter XXI

* **accede** *v.*: agree to; give in

salubrious *adj.*: healthful; wholesome

bathos *n.*: depth

contrarIety *n.*: disagreement

* **incorporeal** *adj.*: without material body or substance

Chapter XXII

Michaelmas *n.*: feast of St. Michael the Archangel, celebrated on September 29

diurnal *adj.*: daily

sackless *adj.*: (Scottish) weak; lacking energy or spirit

Slough of Despond *n.*: deep, hopeless despair; an allusion to a deep swamp in John Bunyan's allegorical tale, *Pilgrim's Progress* (1678).

bugbear *n.*: imaginary goblin used to frighten children into good behavior

* **credulity** *n.*: tendency to believe too readily; lack of doubt

Chapter XXIII

elysium *n.*: complete happiness; paradise

* **odious** *adj.*: disgusting; hateful

Chapter XXIV

interdict *n.*: prohibition

Chapter XXVI

* **lethargy** *n.*: great lack of energy; sluggishness

Chapter XXVII

* **enigmatical** *adj.*: baffling; perplexing

* **attenuated** *adj.*: thin and weak

ling *n.*: heather

vivisection *n.*: surgical operations or experiments performed on living animals

spleen *n.*: bad temper

cockatrice *n.*: in legend, a serpent having the power to kill with a glance

eft *n.*: type of salamander; newt

Chapter XXVIII

bevy *n.*: group

Chapter XXIX

supplications *n.*: earnest requests; prayers

Chapter XXXI

* **adroitly** *adv.*: cleverly; expertly

emulous *adj.*: desiring to equal

appropriate *v.*: take for one's own use, especially without permission

conflagration *n.*: large, destructive fire

saturnine *adj.*: gloomy and silent

Chapter XXXII

hostler *n.*: person who takes care of horses at an inn

stocks *n.*: flowering plants of the mustard family

* **indolence** *n.*: laziness; idleness

automatons *n.*: persons acting automatically, as if they were machines

* **obdurate** *adj.*: hardhearted; stubborn

* **paragon** *n.*: model of excellence

Chapter XXXIII

* **sidled** *v.*: moved sideways

Chapter XXXIV

* **admonition** *n.*: warning to correct a fault

* **protracted** *adj.*: prolonged

Making Meanings: Chapters I–III

Wuthering Heights

First Thoughts

1. With which of the characters at Wuthering Heights do you feel the most sympathy? Explain.

Shaping Interpretations

2. A **first-person narrator** can remain outside the action or be directly involved in the action. Which kind of first-person narrator is Lockwood? How reliable do you think he is? Cite examples from the novel to support your view.

3. How do the details provided by the narrator of Heathcliff's appearance and personality shape your opinion of him?

4. In Chapter I, "Wuthering" is defined as a local word representing turbulence in stormy weather. What details reflect turbulence both inside and outside the house?

5. **Verbal irony** occurs when a speaker says one thing and means something different. Find at least three instances of Lockwood's use of verbal irony.

6. **Suspense** is created by leaving certain questions unanswered until later in the plot. List five questions unanswered thus far about the characters and events that the narrator has witnessed at Wuthering Heights.

7. Though Joseph professes to be a devout Christian, what actions show him to be hypocritical?

8. Summarize Lockwood's two nightmares. Does he believe both are nightmares? Support your view with the text.

READING CHECK

a00.0. 0What is Lockwood's relationship to Heathcliff?

b. Why has Lockwood come to live at Thrushcross Grange?

c. What does Lockwood learn is the relationship between Heathcliff and the two young people?

d. Why must Lockwood spend the night at Wuthering Heights?

e. What does Lockwood learn about Heathcliff's childhood from the diary he finds?

Connecting with the Text

9. What are some reasons that a character in other stories, movies, or TV shows haunts the living?

10. Lockwood draws conclusions about his landlord based on Heathcliff's hospitality. What qualities of a good host does Heathcliff possess or lack? How do you measure hospitality when you are a visitor? Is hospitality still a valued aspect of our culture?

Writing Opportunity

Write a paragraph explaining the **symbolism** of the manor house named Wuthering Heights.

Name _____ Date _____

Reading Strategies: Chapters I–III

Wuthering Heights

Characterization

Complex characterization enhances a story. Complex characters, like people you encounter in life, are unpredictable and multi-dimensional. Brontë's skill in creating complex characters is evident in the opening chapters.

Consider the actions and words of the following characters. Identify dominant traits they possess and then traits that are contradictory.

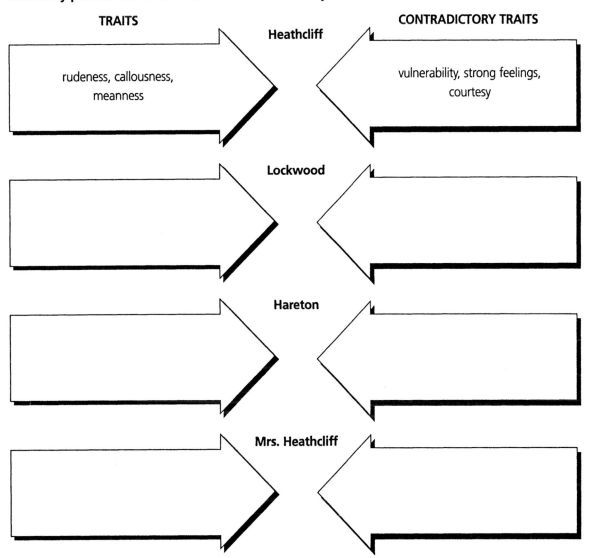

TRAITS → Heathcliff ← CONTRADICTORY TRAITS
- rudeness, callousness, meanness
- vulnerability, strong feelings, courtesy

Lockwood

Hareton

Mrs. Heathcliff

FOLLOW-UP: Based on what you know about these characters so far, compose a paragraph from the point of view of Heathcliff, Hareton, or Mrs. Heathcliff describing Mr. Lockwood.

Novel Notes

Issue 1

Chapters I–III, *Wuthering Heights*

The Story BEHIND

Tea for You and Tea for Me

Afternoon tea, which included food such as small sandwiches, fruit tarts, and assorted cakes, began in the 1840s. Afternoon tea was served between three and six o'clock, and women often slipped into something more comfortable—the loose-fitting "tea gown"—for the ritual. The East India Company imported tea from China, which was the source of 85 percent of the tea drunk in England until 1875. The demand for tea was so great that there was a market for "secondhand tea." In the 1840s, eight factories in London were recycling old tea leaves to resell to thirsty Britons.

FOR YOUR READER'S LOG

Besides influencing your wardrobe, how does the climate in which you live affect your life?

Clothes Closet

FOUL WEATHER GEAR

Walking between Wuthering Heights and Thrushcross Grange in bad weather necessitates the proper clothes. Heathcliff and Mr. Lockwood must have stocked up on the following:

- Gaiters, which protected the lower parts of men's legs from mud, dirt, and rain, were usually made of leather and buttoned up the side. Short gaiters were called "spatterdashes," or "spats."
- Breeches (later known as pants) came down to just below the knees and were made of leather, corduroy, and linen. They were worn with stockings.
- Baggy shirts, called smocks, were worn over breeches.
- Heavy leather shoes completed the ensemble.

Crazy Caricatures

Mr. Lockwood discovers Catherine's caricature of Joseph inside her book. Caricatures, funny drawings that resemble their subjects but exaggerate certain features (like the nose) to make their subjects look ridiculous, were very popular in England during the nineteenth century. Magazines like *Punch* used caricatures to give their readers a more informal look at famous people, such as Queen Victoria.

INVESTIGATE
- Where do you see caricatures today? Find a caricature, bring it to class, and explain the humor in the exaggerated features.

Wuthering Heights

Choices: Chapters I–III

Wuthering Heights

Building Your Portfolio

PERFORMANCE

Tableau

In a group of five or more, choose a scene from *Wuthering Heights*. Assign the roles of characters to group members. Then, arrange yourselves as if you were going to act out the scene, but instead, "freeze" it and ask class members to choose one group member at a time. When chosen, each member "unfreezes" and explains what he or she is thinking during the scene.

ART

From Verbal to Visual

The room where Lockwood spends the night originally belonged to Catherine Earnshaw. Read closely the description in Chapter III and draw either a picture or room plan that locates the chair, clothes-press, oak case, couch, window, and window ledge.

READING STRATEGIES

Dream Weaver

Discuss with a partner the events or discoveries that may have triggered scenes in Lockwood's dreams. Record your thought process in a two-column chart.

- In the first column list events or discoveries.
- In the second column list dream elements that might have resulted from them.

CREATIVE WRITING

Couch Session

A vignette is a short sketch or a memorable scene. Write a vignette of a session between Lockwood and his doctor. Create dialogue in which Lockwood relates his two dreams and the doctor provides him with interpretation and advice. If possible, you and a classmate perform the skit for the class.

Consider This . . .

[Hareton Earnshaw's] thick brown curls were rough and uncultivated . . . his hands were embrowned like those of a common laborer: still his bearing was free, almost haughty . . .

First impressions are often based on appearance and conduct, yet they can sometimes be incorrect or misleading. Discuss the conclusions about Hareton that Lockwood drew from his appearance.

Writing Follow-up: Personal

Compose a two- to four-paragraph personal reflection on an incident when you prematurely drew a conclusion about someone, or when you realized someone's first impression of you was shortsighted.

Novel Notes

Create an activity based on **Novel Notes, Issue 1.** Here are two suggestions:

- Find a caricature in a newspaper or magazine. Write two paragraphs describing what message the caricature is trying to express and how it accomplishes that aim.
- Research the various types of teatimes in British tradition: When is teatime a meal and when is it an afternoon break, for example?

Study Guide | 45

Making Meanings: Chapters IV–IX

Wuthering Heights

First Thoughts
1. What is your opinion of Cathy's decision to marry Edgar?

Shaping Interpretations
2. What elements in the introduction of Hindley's wife, Frances, **foreshadow** her death?
3. Review Heathcliff's description of Thrushcross Grange to Nelly. How does the Grange differ from Wuthering Heights, and how do the houses reflect the personalities of their inhabitants?
4. Explain how Hindley's actions on Christmas Eve and Christmas Day affect Heathcliff.
5. A **foil** is a character whose traits are contrasted with those of another character. Who is Heathcliff's foil in this set of chapters? Explain your answer.
6. What motivates the struggle between Hindley and Heathcliff?
7. After Heathcliff vanishes, a violent thunderstorm splits a tree and damages Wuthering Heights. Explain the possible **symbolic** significance of this event.

READING CHECK
a. In Chapter IV, who takes over the narration?
b. What circumstances first bring Heathcliff to the Heights?
c. How is young Heathcliff treated by individual members of the Earnshaw family?
d. What happens when Heathcliff and Catherine spy on the Lintons?
e. How does Catherine change after five weeks at Thrushcross Grange?
f. How does Heathcliff unwittingly thwart his greatest chance for revenge on Hindley?
g. Why does Catherine choose Edgar over Heathcliff?

Writing Opportunity
Present this information in a paragraph that compares and contrasts the two characters. Support with quotations from the novel.

Extending the Text
8. "Nelly, I am Heathcliff! He's always, always in my mind . . . as my own being. So don't talk of our separation again." Does this express what continues to be the romantic expectation of love? Explain your response and the experiences—personal, fictional, or societal—that led you to this conclusion.

Challenging the Text
9. Some critics criticize Emily Brontë for creating unrealistic characters. Discuss whether Catherine, Heathcliff, and Hindley are realistic or not, and what details make them so.
10. Nelly is opinionated and admits she is not fond of Catherine. How does this affect your view of her opinions about the novel's characters? Is it fair for the author to give her narrator such authority, which may sway the reader's opinion?

Name _____ Date _____

Reading Strategies: Chapters IV–IX

Wuthering Heights

Cause and Effect

When Catherine tells Nelly she intends to marry Edgar, neither of them realizes that Heathcliff is listening in another part of the room. This conversation begins a series of events with tragic consequences for Edgar's parents.

Trace the effects of this overheard conversation. List events that follow chronologically and identify the final, tragic outcome.

| Heathcliff overhears Catherine telling Nelly that she is going to marry Edgar. |

This causes . . .

| |

This causes . . .

| Catherine to stay outside in a violent rainstorm. |

This causes . . .

| |

This causes . . .

| Catherine to be taken to Thrushcross Grange. |

This causes . . .

| |

FINAL OUTCOME

| |

FOLLOW-UP: Compose a paragraph that discusses how Catherine and Heathcliff react to the loss of each other and how their behavior affects others in a harmful way.

Study Guide | 47

Novel Notes

Issue 2

Chapters IV–IX, *Wuthering Heights*

MEDICINE CABINET
Moor Mortality

If an apple a day keeps the doctor away, nineteenth century Britons must have eaten bushels because most lived their entire lives with no formal medical treatment. However, infant mortality rates worsened during the century. Without the inoculations available today, many children never reached adulthood. The big killer of adults and children alike was tuberculosis (consumption). One in six deaths during the nineteenth century was from tuberculosis—more than from scarlet fever, whooping cough, smallpox, measles, and typhus combined.

FOR YOUR READER'S LOG

Heathcliff struggles against others' perception of him based on his appearance. How is this a universal experience?

What's Cookin'?

Nelly loves Christmas and the "rich scent of the heating spices." Popular Christmas fare during Queen Victoria's reign included rich cakes made with fruit, mince pies, turkey, and Christmas pudding flaming with brandy and topped off with a sprig of holly.

On Christmas Eve, a yule log (the largest log that could be found) was hauled home and placed on the hearth to burn through the night. Usually, the log was lit from a burning piece of last year's log (saved for the occasion), and it was important that the log should burn from before sunset until after sunrise.

History in a Nutshell: *Gypsies*

Heathcliff was called a "gipsy brat," because his dark skin resembled that of the Gypsies, a people who have frequently been treated as outcasts.

Here are some interesting facts about Gypsies:
- Gypsies, dark-skinned nomadic people, first came to England around 1490.
- They originated in India, migrating first to Persia and then to Europe.
- Gypsies were first thought to be from Egypt and thus called "Egyptians," which gradually became "Gyptians," and then "Gypsies."
- By one account, there were 10,000 Gypsies in the British Isles by 1528.

INVESTIGATE • *What is the history of Gypsies in the United States?*

The Word PLACE
Poor Little Bird

Hareton is described as being cast out into the world like an "unfledged dunnock." A dunnock is a bird called a hedge sparrow. A common species of hedge sparrow, *prunella modularis,* is found throughout warm regions of Europe. It is a chunky gray and brown bird, about five and one-half inches long, with a slender bill. The hedge sparrow sometimes nests in the abandoned nests of other birds, and both parents take care of the young dunnocks.

Name _____ Date _____

Choices: Chapters IV–IX

Wuthering Heights

Building Your Portfolio

PERFORMANCE

Tabloid TV

In a group of five or six, brainstorm possibilities of a tabloid talk show episode with characters and issues from Chapters IV–IX. The main characters are the guests. There will need to be a host. Confine yourself to information known thus far, though allusions may be made to possible future events.

READING STRATEGIES

Which Suitor Is More Suited?

Consider Cathy's two suitors: Heathcliff and Edgar. Which suitor is more suited to Ms. Earnshaw? You and a partner create a chart to sort out her dilemma.

- In the first column, list at least two reasons that Heathcliff is the more worthy admirer.
- In the second column, list at least two reasons that Edgar is the more worthy admirer.
- In the third column, list reasons Cathy would reject a proposal of marriage from Heathcliff.
- In the fourth column, list reasons Cathy would reject a proposal of marriage from Edgar.

With all this in mind, compose a marriage proposal for each suitor. Present the proposals to the class and allow them to vote on the most persuasive offer.

CREATIVE WRITING

Undying Love

Compose a poem from Catherine to Heathcliff—one that conveys her undying love for him but also explains her choice to marry Linton. The poem should be at least three stanzas in length. Consider imitating Catherine's figurative language, which often contains comparisons with nature.

ART

Comic Strip

Write and illustrate a four-panel (or longer) comic strip based on a scene in this part of the novel. It does not have to be humorous. The tone of your strip should be appropriate to the tone of the book.

Consider This . . .
"Proud people breed sad sorrows for themselves. . . ."

What did Nelly mean by this advice to Heathcliff? How can excessive pride harm a lasting friendship?

Writing Follow-up: Persuasion

With Heathcliff as your audience, compose a two- to four-paragraph persuasive argument in support of Nelly's advice to him. Give at least two reasons for your position; refer to the novel and your own personal history for these supporting examples.

Novel Notes

Create an activity based on **Novel Notes, Issue 2.** Here are two suggestions:

- Research how Gypsies were treated in Britain during this time period.
- What kind of meals were being served at Thrushcross Grange and Wuthering Heights? Research the question, and create sample menus.

Making Meanings: Chapters X–XVII

Wuthering Heights

First Thoughts

1. For which character(s) do you have the most respect? For which do you have the least? Explain.

Shaping Interpretations

2. Why does Nelly not recognize Heathcliff when she first encounters him lurking in the garden of Thrushcross Grange?

3. Why does Catherine tell Heathcliff in front of Isabella that Isabella is infatuated with him?

4. After a mortified Isabella rushes away, how does Catherine unwittingly suggest another avenue of revenge to Heathcliff?

5. Briefly describe the encounter between Heathcliff and Edgar in Chapter XI. What effect does the incident have on Catherine? Why does she blame Edgar more than Heathcliff?

6. How does Heathcliff justify terrorizing Isabella after they return from their honeymoon?

Reading Check

a. Describe how Catherine and Edgar treat Heathcliff when he visits them at Thrushcross Grange.

b. What does Catherine tell Isabella to discourage her from falling in love with Heathcliff?

c. Name three ways Heathcliff takes revenge on Hindley.

d. What is Heathcliff's opinion of Isabella?

e. What does Heathcliff pray for after he hears of Catherine's death?

f. Why does Isabella flee Wuthering Heights and where does she eventually settle?

g. Why doesn't Hareton inherit Wuthering Heights after his father's death?

7. In Chapter XV Nelly arranges a meeting between Heathcliff and Catherine. Summarize what happens at the meeting and later that night after Heathcliff leaves.

8. Does Heathcliff destroy Hindley or does Hindley destroy himself? Explain.

Connecting with the Text

9. Compare the personalities of Heathcliff's two foes—Hindley and Edgar—noting especially what Nelly says about them in Chapter XVII. Considering what you know about human nature and the personalities of the characters in the novel, do you think Heathcliff will triumph over Edgar as easily as he did over Hindley? Give reasons to support your opinion.

Challenging the Text

10. Brontë's decision to have her heroine die halfway through the novel was a daring stroke. How has it already proven effective in creating mystery? What effect do you predict it will have on the rest of the novel?

Writing Opportunity

Develop this opinion into a paragraph response. Provide specific examples from the novel to support your stance.

Name _____ Date _____

Reading Strategies: Chapters X–XVII

Wuthering Heights

Summarizing and Drawing Conclusions

Nelly serves as a catalyst to three quarrels, the last of which results in Catherine's locking herself in her room. The quarrels begin when Nelly tells Catherine she has seen Heathcliff and Isabella embracing.

To better understand the day's events, it is important to understand the quarrels. For each quarrel, list each participant's complaints, then summarize how each quarrel leads to the next quarrel or event. Include Nelly's role.

The First Quarrel ➤ The Second Quarrel ➤ The Third Quarrel

Conflicts:
Catherine

Heathcliff

Outcome/Nelly's role:

Conflicts:
Edgar

Catherine

Heathcliff

Outcome/Nelly's role:

Conflicts:
Edgar

Catherine

Outcome/Nelly's role:

FOLLOW-UP: What do the quarrels reveal about Catherine's goals? Edgar's? Write a short paragraph on each.

Study Guide | **51**

Novel Notes

Issue 3

Chapters X–XVII, *Wuthering Heights*

The Story BEHIND

It's a Man's World—Or Is It?

If Catherine and Edgar had no sons, who would inherit Thrushcross Grange? Heathcliff realizes the heir would be Isabella. Customarily, a woman did not inherit, because the line would die out if she did not marry or, if she did marry, the land could pass to someone outside the family. Families in England kept their wealth and status through the years by passing down their estates from son to son. Two methods were used to make certain that the estates stayed intact:

- **Primogeniture** left the land to the eldest son.
- **Entail** put restrictions on what the eldest son could do with the land, which usually meant tying up the property so that the heir got only the income from the land; he could not sell or mortgage it.

FOR YOUR READER'S LOG

What choices and behaviors by the women in the novel do you perceive to be the results of or reactions to a male-dominated society?

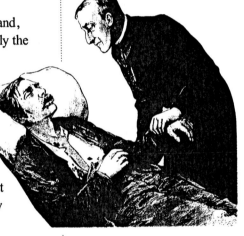

History in a Nutshell: GRAVE CONCERNS

Death occurred all too often in the nineteenth century and was surrounded by customs that might seem strange to us today:

- In the country, the "passing bell" ritual began while the person was on his or her deathbed: the church bell was tolled (rung) six times for a woman and nine for a man, followed by one ring for each year of the dying person's life.
- Professional mourners were hired to dress in black and stand around at the funeral to add dignity to the occasion.
- A suicide, before 1823, was required by law to be buried at a crossroads—to spread the evil in four directions.
- Cremation was not used in England until 1874.

 INVESTIGATE
- *Research death customs around the globe. Then, present an interesting custom to the class.*

The Story BEHIND

Grouse Hunting

Heathcliff sends Mr. Lockwood a brace (pair) of grouse as a gift. Grouse is a small game bird, which was hunted for sport on the moors in northern England and Scotland. In the nineteenth century, grouse season lasted from August 12 through November. Not coincidentally, the start of grouse season was also the time when Parliament (the chief lawmaking body) ended its session and upper-class Britons left the city to hunt on the moors.

Quotation Corner

Perhaps Catherine and Heathcliff should have listened to these words of wisdom:

"These violent delights
have violent ends
And in their triumph die,
like fire and powder,
Which, as they kiss,
consume."

—Shakespeare,
Romeo and Juliet

Name _____ Date _____

Choices: Chapters X–XVII

Wuthering Heights

Building Your Portfolio

PERFORMANCE
"The Play's the Thing"
In a group of five or more, plan and perform a skit based on a scene in this section. You may want to choose a narrator who will give any background necessary for the audience to understand the skit.

VIDEO RECORDING
Television News Magazine
With a partner, choose a character from *Wuthering Heights* and rehearse and videotape a TV interview. One partner takes the role of the interviewer and the other plays the *Wuthering Heights* character. Show the video to the class.

CREATIVE WRITING
Missing Chapter
Write a chapter that fills in the gap from the time Heathcliff vanishes from Wuthering Heights to the time he returns after Catherine and Edgar marry. Be sure to explain how he acquired his money, his changed appearance, and his accent.

CRITICAL WRITING
Dear Emily
Write a letter to Emily Brontë in which you discuss your responses to the plot, characters, and themes of *Wuthering Heights*. Be sure to tell the writer what you most like about the book so far, what confuses or bothers you, and what direction you think the story should take. Since you can't mail your letter, read it to your class or small group.

POETRY
Look what I found
Choose one or two lines from your book that seem particularly descriptive or that contain examples of figurative language that you think are especially vivid. On a blank sheet of unlined paper, write the sentence(s) as they appeared in your book. Next, write the sentences so that the placement of the letters and words suggests the image being described. You may want to illustrate your "found poem."

Consider This . . .
"I gave him my heart, and he took and pinched it to death, and flung it back to me. People feel with their hearts . . . and since he has destroyed mine, I have not power to feel for him."

Can a heart get so injured that it can't feel anymore? Do you think Isabella overreacted to Heathcliff's behavior or took too long to react? Explain.

Writing Follow-up: Cause and Effect
In two to four paragraphs, trace the evolution of Isabella's feelings for Heathcliff. Begin with her jealous attraction and conclude with her cold spitefulness. Explain how Heathcliff's actions caused these emotions.

Novel Notes
Create an activity based on **Novel Notes, Issue 3.** Here are two suggestions:
- Draw and label weapons used to hunt grouse during this time period.
- Research the property laws governing marriage and death current in your locale.

Study Guide | 53

Making Meanings: Chapters XVIII–XXV

Wuthering Heights

First Thoughts

1. What is your reaction to Linton's behavior?

Shaping Interpretations

2. Compare Nelly's opinion of Cathy to the opinion she had of Cathy's mother. How does Nelly's attitude affect your sympathy for Cathy?

3. Why does Hareton curse Cathy during their first meeting, and, **ironically,** what does she then learn about him?

4. What details contribute to the reader's feeling of dread for Linton when he is taken to Wuthering Heights to live with his father?

5. On Cathy's visit to Wuthering Heights to meet Linton, who shows her around the stables? Why?

6. Explain how Cathy denies her true nature by preferring Linton over Hareton. In what way do her actions recall those of her mother?

7. How does Heathcliff convince Cathy to visit Wuthering Heights again, and how does she then convince Nelly to let her go?

8. Why does Linton turn away when Cathy tries to kiss him? What does this reveal about his personality?

Connecting with the Text

9. Linton seems to have few redeeming qualities; however, in Chapter XXIV he tries to explain to Cathy why he is "bad in temper, and bad in spirit." Do you find Linton's explanation of Heathcliff's treatment of him adequate reason for his selfishness? Explain why doubt of one's own worth might make one "cross and bitter."

10. Verbal communication plays a large role in *Wuthering Heights*. Significant events occur because a character can or cannot read, write letters, or express his or her feelings. Do you think the ability to read and write has the same power today that it did in Brontë's time? Explain.

READING CHECK

a. How do Cathy and Hareton initially react to each other when they meet?

b. Explain how Cathy, Hareton, and Linton are related.

c. How long does Linton stay at Thrushcross Grange?

d. When do Linton and Cathy meet again?

e. What is Heathcliff's plan for Linton and Cathy?

f. Why does Cathy write to Linton?

g. What circumstances provide Cathy with the opportunity to visit Linton frequently?

Writing Opportunity

Explain how Cathy's situation and choices differ from those of her mother.

Name _____ Date _____

Reading Strategies: Chapters XVIII–XXV

Wuthering Heights

Comparing

In looks and personality, Cathy shares traits with both her parents.

In the graphic below, where Cathy's circle overlaps her father's oval, list the traits and appearance they share. Do the same for the overlapping space of Cathy and her mother.

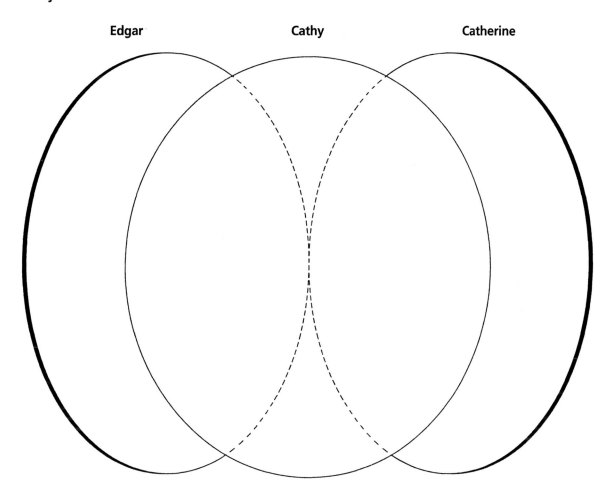

Edgar   Cathy   Catherine

FOLLOW-UP: Describe one of Cathy's reactions to an event. Then, explain the personality trait behind the reaction and whether the trait is shared by Cathy's mother or father.

_____

_____

_____

Study Guide | **55**

Novel Notes

Issue 4

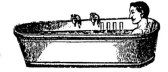

Chapters XVIII–XXV, *Wuthering Heights*

The Story BEHIND

Bathing Beauties

Victorians did not have hot daily showers, dry cleaners on every corner, or much in the way of deodorants. How did they stand being near each other?

The only body parts washed regularly in England during the early nineteenth century were the hands, neck, and arms. By the 1860s, the middle class took one bath a week—on Saturday night. The entire family bathed on the same night, because it was such a bother to heat all the water. At about this time, the well-to-do began to install special rooms for baths, and they may have bathed more frequently.

Part of the problem with bathing regularly was not having water on hand. Water had to be brought from streams, rivers, or wells in the country and from the town pump in villages.

It wasn't easy to wash clothes either. Some washed them in a river or stream and beat the clothes with a paddle until clean. Soap was a problem: it had to be made from tallow (animal fat), which was often saved for food; soap was taxed if not homemade, so it was expensive; and soap required hot water, not always easily accessible. All and all, laundry day was such a chore that many households who could afford it hired a washerwoman for the task.

INVESTIGATE • *Who invented indoor plumbing, and when did it come into common use?*

FOR YOUR READER'S LOG
What is the significance of nicknames and endearments for you? Can you understand Edgar's decision not to call his wife Cathy? Explain.

The Word PLACE

A Term of Endearment?
Heathcliff calls Linton his "puling chicken." *Puling* refers to whining or to crying in a weak voice. It comes from the verb *pule*, believed to have originated from the French *piauler*, "peep" or "whine".

56 | *Wuthering Heights*

Name _____ Date _____

Choices: Chapters XVIII–XXV

Wuthering Heights

Building Your Portfolio

PERFORMANCE

Round Table

Divide into groups of five, with group members taking the roles of different characters from the novel. *Keeping in character,* have a round table discussion on issues touched upon in the novel and still relevant today (for example, status of the wife in the home, young marriages, overprotective parents). You may wish to appoint a moderator to assure that members stay in character.

CREATIVE WRITING

Rhyme with a Reason

Compose a poem of at least twelve lines inspired by the events, atmosphere, or characters in this section. You can write from your own point of view or as if you were a character.

PERFORMANCE

Writing: Radio Script

With a partner, prepare a script for a call-in radio show. Callers are characters from *Wuthering Heights* seeking advice about relationships. Have three to five characters call in. Present the script in class with one person playing the radio host, the others playing the advice-seekers.

ART

A Picture Is Worth a Thousand Words

As young children, we all enjoyed the illustrations in books. They made the story come alive for us. Create an illustration of an event in this section of the novel. Work on paper that is of the same dimensions as the pages of your book. Provide the passage to which the picture corresponds.

Consider This . . .

"You are so much happier than I am . . . Papa talks enough of my defects, and shows enough scorn of me, to make it natural I should doubt myself. I doubt whether I am not altogether as worthless as he calls me."

How important are positive comments from friends and family in how we act and feel? Consider this from a personal experience or personal knowledge.

Writing Follow-up: Compare and Contrast

In a two- to four-paragraph essay, compare and contrast the way Linton's father treats him to the way Cathy's father treats her and the effect this interaction with fathers has had on the children's personalities.

Novel Notes

Create an activity based on **Novel Notes, Issue 4.** Here are two suggestions:

- Research the process for making homemade soap with animal fat.
- Investigate other aspects of everyday life in the first half of the ninteenth century.

Making Meanings: Chapters XXVI–XXXIV

Wuthering Heights

First Thoughts
1. By the novel's conclusion, what is your feeling toward Heathcliff?

Shaping Interpretations
2. What do the events surrounding the marriage of Cathy and Linton (Chapter XXVII) reveal about the characters of those involved?
3. Why does Heathcliff detain Cathy at Wuthering Heights for so long?
4. What final effort does Edgar make to keep Cathy's personal fortune away from Heathcliff, and how does Heathcliff thwart him?
5. What does Heathcliff tell Nelly he has twice attempted to do at Catherine's grave? How did Catherine's "presence" affect him?
6. When Heathcliff sees Cathy teaching Hareton to read, why does he not forbid their interaction? Explain the "strange change" that Heathcliff feels approaching.
7. Identify three or more ways in which Cathy openly defies Heathcliff. Why do you suppose he is unable to subdue her as he did her two cousins, Linton and Hareton?
8. In previous sections of the novel, Heathcliff was engaged in **external conflicts** with Hindley and Edgar. In this final section, his external conflict primarily involves Cathy. Explain how losing his struggle with Cathy helps him resolve the lifelong **internal conflict** that has fueled his revenge.
9. When Lockwood returns to Wuthering Heights, he finds blooming flowers and a neatly tended garden. **Symbolically,** what does this signify?

Connecting with the Text
10. "[Hareton] said he wouldn't suffer a word to be uttered, in his disparagement. . . ." Explain Hareton's loyalty to Heathcliff. Base your explanation on novel details and your experience or observations of family dynamics.

Challenging the Text
11. The author uses multiple narrators. Could any of the narrators have been eliminated? If so, how would this change the story?

Reading Check
a. How does Linton appear when Nelly and Cathy meet him for a ride across the moors?
b. How does Cathy come to marry Linton?
c. How does Linton help Cathy escape to return to her father?
d. Of what two things does Heathcliff say Hareton reminds him?
e. According to Zillah, how did Cathy get along with Hareton after Linton's death?
f. How does Heathcliff respond when Nelly tells him he should repent his injustices?
g. What does Lockwood discover when he returns to Wuthering Heights after having been away nine months?

Writing Opportunity
Analyze how the marriage between Cathy and Hareton represents a resolution of conflicts caused by the failure of Catherine and Heathcliff to marry each other.

Name _____ Date _____

Reading Strategies: Chapters XXVI–XXXIV

Wuthering Heights

Problem and Solution

Heathcliff plans carefully to secure Thrushcross Grange as part of his scheme for revenge against Edgar.

In the graphic organizer below, list each circumstance or obstacle that stands in the way of Heathcliff's acquiring the Grange and his means of overcoming that obstacle. Structure your responses like the example (drawn from circumstances earlier in the novel).

Circumstance/Obstacle		
Heathcliff wants one of Hindley's colts.	→ How Heathcliff overcomes it	Heathcliff allows Hindley to beat him physically, and then threatens to tell Mr. Earnshaw unless Hindley gives him the colt.
Circumstance/Obstacle	→ How Heathcliff overcomes it	
Circumstance/Obstacle	→ How Heathcliff overcomes it	
Circumstance/Obstacle	→ How Heathcliff overcomes it	

FOLLOW-UP: Compose a paragraph that analyzes Heathcliff's character, based on how he goes about acquiring what he wants.

Novel Notes

Issue 5

Chapters XXVI–XXXIV, *Wuthering Heights*

The Story BEHIND
LEGAL EAGLES

Emily Brontë refers to Mr. Green as a "lawyer," but there were several titles for lawyers in England during that time and still are today. Then, lawyers were divided into two groups: those who argued before the court and those who handled other legal matters.

- **Barristers** argued cases in the Court of Chancery and enjoyed the highest prestige of any lawyers by the mid-1800s.
- **Advocates** argued in the admiralty and church courts.
- **Solicitors** handled legal problems not requiring court appearance, such as wills and estates, and hired barristers to represent their clients before the courts.
- **Proctors** assisted the advocates.

INVESTIGATE
- *How did these divisions change after an 1873 court reorganization?*

FOR YOUR READER'S LOG
What role do lawyers play in society today?

What's Cookin'?

Cathy places primroses in Hareton's plate of porridge. **Porridge,** primarily a British dish, is a soft food made from cooking cereal such as oatmeal or meal with milk or water until it thickens. Porridge was often served in a small metal vessel called a porringer.

The Story BEHIND
Here Comes the Bride . . .

Victorians were allowed by law to marry a cousin (even a first cousin), as Cathy does twice, but were not allowed to marry sisters- or brothers-in-law. Below are some other marriage customs.

- A boy could marry without his parents' consent at age 14 and a girl at 12.
- With marriage, a woman's personal property became her husband's to do with as he wished, but he could not sell or mortgage her land, only collect rent from it.
- A wife could not make a contract on her own, sue, or make a will without her husband's consent.
- A husband was liable by law for his wife; if she committed a crime, it was assumed that she was acting under his influence.
- In the late eighteenth century, names of aristocratic families were often hyphenated to include the wife's family's name.

Name _____ Date _____

Choices: Chapters XXVI–XXXIV

Wuthering Heights

Building Your Portfolio

PERFORMANCE/MEDIA

Newscast
In a group of three to five, write and present a newscast on the events that have been occurring in and around Wuthering Heights and Thrushcross Grange. Include a news story, a short feature, weather, and sports (perhaps a hunting report). Present to the class.

CREATIVE WRITING

Let's Talk
With a partner, rewrite a dialogue involving two characters. Place the scene in a different setting (time, place, or both), but stay true to the characters' personalities. Present the dialogue in class.

ART

Collage
Choose an element of *Wuthering Heights* from the following list: characters, setting, themes, language, events. Using words and pictures cut out of old magazines or newspapers or downloaded off the Internet, as well as other supplies, create a collage illustrating the element you have chosen.

READING STRATEGIES

Time Line
The events of these last chapters cover several months and are recounted in both present tense by Lockwood and flashback by Nelly. Create a time line to straighten out the sequence of events that brings this novel to a close.

WRITING/PERFORMANCE

Eulogy
Despite Cathy's dislike of Heathcliff, Hareton would not speak ill of the man who had raised him and was the only one to mourn his death. Imagine that Hareton is asked to write and deliver a eulogy, a speech praising a person after death, for Heathcliff. What positive points of Heathcliff's personality might he praise? How might he explain his loyalty and love for this man who had harmed so many people? Compose a eulogy such as Hareton would write and deliver this speech to the class.

Consider This . . .
He had a fixed idea . . . that, as his nephew resembled him in person, he would resemble him in mind.

How frequently do we make the mistake of equating looks with behavior? Think of a time when you anticipated a certain behavior because of an individual's appearance.

Writing Follow-up: Reflection

Compose a two- to four-paragraph reflection on the idea of equating looks with behavior. Refer to the novel and personal history or world/national history.

Novel Notes

Create an activity based on **Novel Notes, Issue 5.** Here are two suggestions.

- Research first-cousin marriages, typical when the book was written. When and why did the custom change?
- Research the history of the popular fashion of wigs that were worn in court.

Name _____ Date _____

Novel Review

Wuthering Heights

MAJOR CHARACTERS

Use the chart below to keep track of the characters in this book. Each time you come across a new character, write the character's name and the number of the page on which the character first appears. Then, jot down a brief description. Add information about the characters as you read. Put a star next to the name of each main character.

NAME OF CHARACTER	DESCRIPTION

FOLLOW-UP: A *dynamic character* changes in some important way as a result of the story's action. In a paragraph, trace the transformation of one dynamic character from the time the character is introduced through the conclusion of the novel.

Name _____ Date _____

Novel Review (cont.)

Wuthering Heights

SETTING

Time _____

Most important place(s) _____

_____

One effect of setting on plot, theme, or character _____

_____

_____

PLOT

List key events from the novel.

- _____   • _____
- _____   • _____
- _____   • _____

Use your list to identify the plot elements below. Add other events as necessary.

Major conflict / problem _____

_____

Turning point / climax _____

_____

Resolution / denouement _____

_____

MAJOR THEMES

- _____
- _____
- _____

Study Guide | **63**

Name _____ Date _____

Literary Elements Worksheet 1

Wuthering Heights

Symbolism

In addition to using Wuthering Heights and Thrushcross Grange as symbols, Emily Brontë uses the weather and nature to signify the emotions and conflicts her characters experience.

In the graphic organizer below, list three examples of symbolism in weather or nature. Write in the passage from the novel, identify the conflict or emotion being represented, and identify the chapter and page.

Symbol	Conflict or emotion

Location in text: Chapter ___, page ___

Symbol	Conflict or emotion

Location in text: Chapter ___, page ___

Symbol	Conflict or emotion

Location in text: Chapter ___, page ___

FOLLOW-UP: Try using a setting or another type of symbol to represent a mood or emotion you have felt. Write at least a paragraph.

Name _____ Date _____

Literary Elements Worksheet 2

Wuthering Heights

Multiple Narrators

In *Wuthering Heights,* Emily Brontë uses multiple first-person narrators; this technique creates immediacy and allows the reader to experience the story as it unfolds. It also shows different points of view toward the characters and different attitudes toward the action.

In the graphic organizer below, identify the narrators for each section of the novel. Then explain how the narrators affect the way you, the reader, interpret the events they describe.

Chapters	Narrators	Effect of narrator(s)
Chapters I–III		
Chapters IV–IX		
Chapters X–XVII		
Chapters XVIII–XXV		
Chapters XXVI–XXXIV		

FOLLOW-UP: Compose a paragraph that discusses the reliability of three of these narrators.

Study Guide | **65**

Name _____ Date _____

Literary Elements Worksheet 3

Wuthering Heights

Indirect Characterization

Often characters in *Wuthering Heights* are developed through **indirect characterization.** Instead of describing them directly, Brontë reveals her characters as they are observed by one another.

In the boxes on the left, list examples of Heathcliff's words and actions. Then, in the boxes on the right, list words and actions that show the interaction of the four women with Heathcliff. Finally, list Heathcliff's character traits as they are revealed by these examples.

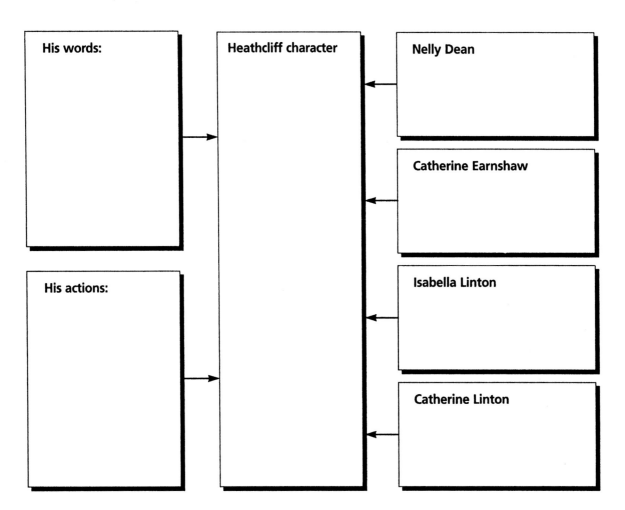

FOLLOW-UP
- Formulate the above information about Heathcliff's character as revealed by indirect characterization into a brief essay of one or two paragraphs.
- Consider the interaction between two of the female characters mentioned above. In a paragraph explain what their words about and actions toward each other reveal about them.

66 | *Wuthering Heights*

Name _____ Date _____

Literary Elements Worksheet 4

Wuthering Heights

Foreshadowing

Brontë frequently uses Nelly to foreshadow or hint at what will happen later in the novel. This technique creates suspense in Nelly's listener, Lockwood, and in the reader.

Below are some quotations from Nelly that foreshadow later events. In the boxes on the right, summarize the event that each quotation foreshadows.

> I had a presentiment in my heart, that he had better have remained away.

1.

> . . . for the space of half a year the gunpowder lay as harmless as sand, because no fire came near to explode it.

2.

> I thought she would scarcely venture forth alone . . . Unluckily, my confidence proved misplaced.

3.

> Poor thing! I never considered what she did with herself after tea.

4.

FOLLOW-UP: Choose one of the above quotations and in a paragraph explain how it foreshadows the events you identified. Be sure to include the circumstances surrounding the quotation and the events.

Study Guide | 67

Name _____ Date _____

Literary Elements Worksheet 5

Wuthering Heights

Gothic Literary Elements

Lockwood's night at Wuthering Heights (Chapter III) introduces Gothic elements into what has been up to that point a very realistic story.

In the graphic organizer below, identify at least three other scenes that include Gothic elements, then identify the elements and their location in the text.

Scene	Gothic elements

Location in text: Chapter ___, page ___

Scene	Gothic elements

Location in text: Chapter ___, page ___

Scene	Gothic elements

Location in text: Chapter ___, page ___

FOLLOW-UP: How does the Gothic aspect of the story contribute to your enjoyment of *Wuthering Heights*? In a paragraph, explain.

Name _____ Date _____

Vocabulary Worksheet 1 — Chapters I–IX

Wuthering Heights

A. Circle the letter of the word or phrase that best defines the italicized word in each quotation from *Wuthering Heights*.

1. "The Lord help us!" he soliloquised in an undertone of *peevish* displeasure. . . .
    - a. barely noticeable
    - b. impish
    - c. unintentional
    - d. ill-tempered

2. He is a dark-skinned gypsy in aspect . . . rather *slovenly*, perhaps, yet not looking amiss with his negligence. . . .
    - a. untidy
    - b. attractive
    - c. wrinkled
    - d. mean-spirited

3. I sat still; but, imagining they would scarcely understand *tacit* insults, I unfortunately indulged in winking and making faces. . . .
    - a. witty
    - b. tasteless
    - c. silent
    - d. outspoken

4. I must trust to my own *sagacity*.
    - a. hasty guessing
    - b. sound judgment
    - c. old age
    - d. sense of humor

5. [S]till it wailed, "Let me in!" and maintained its *tenacious* grip, almost maddening me with fear.
    - a. tentative
    - b. angry
    - c. painful
    - d. firm

6. "What *culpable* carelessness in her brother!" exclaimed Mr. Linton, turning from me to Catherine.
    - a. ignorant
    - b. blameworthy
    - c. willful
    - d. absurd

7. The little party recovered its *equanimity* at sight of the fragrant feast.
    - a. composure
    - b. high spirits
    - c. equality
    - d. optimism

8. [H]er husband persisted *doggedly*, nay, furiously, in affirming her health improved every day.
    - a. angrily
    - b. stubbornly
    - c. in a frantic way
    - d. gloomily

9. Linton evinced disgust and *antipathy* to Heathcliff. . . .
    - a. indifference
    - b. a sick feeling
    - c. deep rage
    - d. strong dislike

10. [H]e has an erect and handsome figure; and rather *morose*. Possibly, some people might suspect him of a degree of underbred pride. . . .
    - a. haughty
    - b. smug
    - c. gloomy
    - d. deliberate

Study Guide | 69

Vocabulary Worksheet 1 (cont.)

Chapters I–IX

Wuthering Heights

11. Mrs. Linton took off the gray cloak of the dairy maid which we had borrowed for our excursion, shaking her head and *expostulating* with her. . . .
    - a. consulting seriously
    - b. agreeing enthusiastically
    - c. arguing vehemently
    - d. laughing playfully

12. [T]ake him . . . you beggarly *interloper!* and wheedle my father out of all he has. . . .
    - a. thief
    - b. instigator
    - c. fool
    - d. intruder

13. "I hate you to be fidgeting in my presence," exclaimed the young lady *imperiously*. . . .
    - a. playfully
    - b. arrogantly
    - c. anxiously
    - d. timidly

14. He drew back in *consternation*.
    - a. reckless abandon
    - b. bewildered fear
    - c. unabashed pleasure
    - d. thoughtful hesitation

15. The little witch put a mock *malignity* into her beautiful eyes. . . .
    - a. ill will
    - b. humor
    - c. fear
    - d. seriousness

B. Match each word in the left-hand column with its correct meaning from the right-hand column.

_____ 16. stalwart   a. bully

_____ 17. querulous   b. robust

_____ 18. dour   c. gloomy

_____ 19. winsome   d. dull

_____ 20. munificent   e. complaining

_____ 21. vapid   f. imperceptible to the touch

_____ 22. hector   g. generous

_____ 23. manifested   h. charming

_____ 24. impalpable   i. irritably

_____ 25. petulantly   j. revealed

Name _____ Date _____

Vocabulary Worksheet 2 — Chapters X–XXXIV

Wuthering Heights

A. Circle the letter of the word or phrase that best defines the italicized word in each quotation from *Wuthering Heights*.

1. Joseph . . . confined his feelings regarding him to muttered *innuendoes*. . . .
   - a. indirect remarks
   - b. lies
   - c. curses
   - d. praises

2. [H]e put his fingers to his eyes to remove *incipient* tears.
   - a. embarrassing
   - b. ample
   - c. effeminate
   - d. beginning

3. [H]e clung to me with growing *trepidation*. . . .
   - a. affection
   - b. strength
   - c. anxiety
   - d. fearlessness

4. I whispered to Catherine that she mustn't on any account *accede* to the proposal.
   - a. agree
   - b. respond
   - c. listen
   - d. disagree

5. What use were anger and protestations against her silly *credulity*?
   - a. stupidity
   - b. deep faith
   - c. lack of doubt
   - d. simplicity

6. That proposal, unexpectedly, roused Linton from his *lethargy*, and threw him into a strange state of agitation.
   - a. fever
   - b. peaceful rest
   - c. meditation
   - d. sluggishness

7. "But my father threatened me," gasped the boy, clasping his *attenuated* fingers.
   - a. outstretched
   - b. weak and thin
   - c. bruised and scratched
   - d. blue and cold

8. I saw she was sorry for his persevering sulkiness and *indolence*. . . .
   - a. idleness
   - b. resentfulness
   - c. childishness
   - d. unhealthiness

9. I didn't know whether it were not a proper opportunity to offer a bit of *admonition* . . . .
   - a. secret information
   - b. gossip
   - c. warning
   - d. encouragement

10. I vainly reminded him of his *protracted* abstinence from food. . . .
    - a. promised
    - b. extended
    - c. foolhardy
    - d. impractical

Study Guide | 71

Name _____ Date _____

Vocabulary Worksheet 2 (cont.)   Chapters X–XXXIV

Wuthering Heights

B. Circle the letter of the *antonym*—the word that is most nearly *opposite* in meaning—for each word in bold type.

11. **sanguine** (a) pessimistic (b) depressed (c) quarrelsome (d) optimistic

12. **allayed** (a) calmed (b) provoked (c) believed (d) assured

13. **incorporeal** (a) unbelievable (b) martial (c) light (d) touchable

14. **enigmatical** (a) mysterious (b) apparent (c) mechanized (d) symbolic

15. **obdurate** (a) rude (b) solid (c) flexible (d) noticeable

C. Match each word in the left-hand column with its correct meaning from the right-hand column.

_____ 16. recommenced   a. summary

_____ 17. usurped   b. arrogant

_____ 18. presumptuous   c. cleverly

_____ 19. inveterate   d. started again

_____ 20. paragon   e. crept sideways

_____ 21. recapitulation   f. habitual

_____ 22. sidled   g. took possession of

_____ 23. incarnate   h. to make flesh

_____ 24. adroitly   i. greed

_____ 25. avarice   j. ideal

72 | *Wuthering Heights*

Novel Projects

Wuthering Heights

Writing About the Novel

DIARY

Private Moments
Write a series of entries for a diary kept by either Catherine, Heathcliff, Edgar, Cathy, or Hareton while he or she falls in love. Considering what you now know of the character, discuss the obstacles to be overcome and the hopes or fears the character might have. *(Creative Writing)*

LITERARY ANALYSIS

Heroic Measures
Research the Romantic figure known as "the Byronic hero," and write an essay comparing him to the character of Heathcliff. Consider their attitudes toward conventional morality, their attitudes toward fate, their emotional states, and even their physical appearance. Does the Byronic hero inspire sympathy, fear, or both? Do you think that any contemporary people fit the Byronic description? Explain. *(Critical Writing)*

SEEING CONNECTIONS

A Room with a View
The image of open windows occurs several times in the book. Lockwood dreams of Catherine's ghost entering by a window; Cathy later escapes from Wuthering Heights through that window in her mother's room; and Heathcliff dies alongside it. In a short essay, explain what you think this **symbol** means and how it adds depth to your understanding of the novel, its **theme,** or its **style.** *(Critical Writing)*

RESPONDING TO A CRITIC

True Love?
The critic Clifford Collins in *Themes and Conventions* described the love of Catherine and Heathcliff as

> a life-force relationship, a principle that is not conditioned by anything but itself. It is a principle because the relationship is of an ideal nature; it does not exist in life. . . . Further, their love is the opposite of love conceived as social and conventional acceptance, the love that Catherine has for Edgar which has only her conscious approval. What she feels for Heathcliff is a powerful undercurrent, an acceptance of identity below the level of consciousness.

In three to five paragraphs, respond to Collins's two assertions: that Catherine and Heathcliff's love is an ideal relationship that does not exist in life, and that their love is opposed to social conventions. Do you agree or disagree? Support your view with reasons and specific evidence from the novel. *(Critical Writing)*

BIOGRAPHICAL SKETCH

Intense, Inspired, Individual
Research the life of Emily Brontë and the Brontë family. You may want to delve especially into the sisters' relationships with their father, their brother, or their publisher. Write a sketch of the author's life in which you blend biographical information and insight into her writing. Highlight connections that you find between the historical facts and *Wuthering Heights*. *(Critical Writing)*

MISSING PERSPECTIVE

The "Other" Man
Select a scene from the novel involving Edgar: the Christmas visit of Chapter VII, his declaration of love to Catherine in Chapter VIII, or his discovery of Catherine in Heathcliff's arms in Chapter XV. Rewrite the scene from Edgar's point of view. Use Brontë's dialogue, but describe the events as Edgar would have seen them. *(Creative Writing)*

Novel Projects (cont.)

Wuthering Heights

Cross-Curricular Connections

MUSIC

Heathcliff, can you hear me?
With eight to ten classmates, select one character in *Wuthering Heights* to be the star and present a rock opera of that character's story. The opera should include six to eight songs. The songs and music can be original, or lyrics to existing songs can be rewritten. Tape and present or perform for the class.

ART/GEOGRAPHY

Where is a heath cliff?
Draw a map of the locale of *Wuthering Heights*, showing with words or symbols where key events took place. You may want to make a topographical map in order to suggest the terrain that plays such a significant role in the atmosphere of the novel.

LANGUAGE ARTS

Woodworking/Home Design
Construct a model of Wuthering Heights or Thrushcross Grange. Supplement details from the novel with research on Yorkshire manor houses of the era. Keep in mind that the Heights was built in the 1500s.

PSYCHOLOGY

Character Analysis
Prepare a psychological profile of at least six characters in *Wuthering Heights* and suggest how their emotional health could be improved. You may want to consider such elements as emotional control, respect for authority, self-confidence, pride, obsessions, and will power.

MATHEMATICS

Checking My Assets
Research and estimate the value of the Wuthering Heights and Thrushcross Grange estates. Consider land, buildings, household goods, farming implements, farm animals, servants, and personal property. Compare the value in 1802 with the value today.

HISTORY

Life in London—Historical Fiction
Write a chapter that reveals what happens to Isabella during her years in London. For accuracy, research what life in London was like during that time period. Consider historic figures and events, writers, clothing, and any other aspects of life.

ART

Marketing a Masterpiece
Design a dust jacket for a new edition of *Wuthering Heights*. Choose carefully the image for the cover: Which character or scene will embody your idea of the novel? Write flap copy that tells just enough of the story to entice a reader, provide information about the author, and perhaps even add a reader's testimonial or two.

HOME ECONOMICS

Literary Cuisine
Research and prepare part of a meal that some of the characters might have enjoyed in *Wuthering Heights*. Remain as close to the authentic recipes as you can. Document the preparation process of the meal with photos or a video.

Novel Projects (cont.)

Wuthering Heights

Multimedia and Internet Connections

NOTE: Check with your teacher about school policies on accessing Internet sites. If a Web site named here is not available, use key words to locate a similar site.

VIDEO: RECORDING

My guests today are . . .

With six to eight classmates, prepare for an Oprah Winfrey–type show. The moderator will interview characters from *Wuthering Heights*. Select a topic that deals with a family problem these characters have experienced. Rehearse and tape or present in class.

COMPUTER: GAME

It's Your Turn

Create a computer game based on the characters and events in *Wuthering Heights*. Include clearly written instructions. Your classmates should learn what happened in the book and be tested on their familiarity with the book by playing your game.

FILM: REVIEW

Two Thumbs Up?

Watch one of the film adaptations of the novel, and compare the movie to the book. How do the characters and settings in the movie differ from what you imagined as you were reading the book? What details were left out, and what details were added? How successful was the filmmaker in presenting the story? What was *not* successful?

VIDEO: MINISERIES

Lights! Camera! Action!

If this novel were a miniseries, what might one episode look like? You and your production crew of classmates are to bring some major scenes from this novel to the small screen. Create dialogue where necessary, along with sets and costumes. Choose a sequence of events in *Wuthering Heights* that will be of high interest to an audience. Videorecord your performance. Include opening and closing credits to give yourselves recognition. Perhaps your class can have an opening gala to premiere your film.

AUDIO: RECORDING

Caught on Tape

With several classmates, create a radio or book-on-tape reading of a section of the novel. Choose an interesting passage of no less than twenty pages. Listen to an old radio show or a book on tape to develop ideas for making your reading engaging to the listener. Consider sound effects, music, and accents.

INTERNET: WEB PAGE

Surf *The Heights*

Design a Web page for *Wuthering Heights*. Your site should familiarize Web browsers with the plot, the setting (time period and locale), and the author of the novel. Include images to make your page more visually exciting, and offer links to other Internet sites that provide supplementary information that could enhance a reader's enjoyment of *Wuthering Heights*. For example, you might include links to pages about London or Liverpool at the turn of the nineteenth century or manor houses of Northern England.

INTERNET: RESEARCH

Moor Tour

Using the Internet, research and plan a one-week trip to Yorkshire County. The week will be spent touring sites relating to Emily Brontë and *Wuthering Heights*. Create a daily travel plan, including tours, meals, and overnight lodging. Use photographs, a map, or graphics downloaded from the Internet. The plan should explain the choices for each day.

Introducing the Connections

The **Connections** that follow this novel in the HRW LIBRARY edition create the opportunity for students to relate the book's themes to other genres, times, and places and to their own lives. The following chart will facilitate your use of these additional works. Succeeding pages offer **Making Meanings** questions to stimulate student response.

Selection	Summary, Connection to Novel
Early Autumn Langston Hughes *short story*	Death ultimately separated Heathcliff and Catherine, but life more frequently divides couples. In this slight, poignant story of a chance meeting, life has changed two people who had once been in love, or has it?
If the Stars Should Fall, Samuel Allen (Paul Vesey) **Sorrow Is the Only Faithful One** Owen Dodson *poems*	Both poems express deeply held and long-lasting grief and sorrow. In both, as in the novel, mourning is mirrored in the imagery of nature.
I see around me tombstones grey Emily Brontë *poem*	Brontë wrote this poem six years before *Wuthering Heights* was published. The initial image is of the tombstones that almost surrounded the Brontë home, and its theme is a passionate love of the earth itself.
"Mr. Bell's" *Wuthering Heights* anonymous *reviews*	Critics who first read the book assumed that only a man would have written it. The first reviewer finds a moral lesson, but the second wishes that the writer had not dragged such "coarse and loathsome" matters into the light.
"Reader, I Married Him" Daniel Pool *essay*	The famous line from *Jane Eyre*, by Emily's sister Charlotte, titles this discussion of marriage customs of the times as exemplified in literary works.

Introducing the Connections *(cont.)*

Selection	Summary, Connection to Novel
Heston Grange from *All Creatures Great and Small* James Herriot *personal essay*	This account of Herriot's visit to a farm in the moors shows that life in some ways had stayed the same there for a hundred years.
The Unquiet Grave anonymous *ballad*	After mourning by her grave for a year, a man hears his dead love's voice tell him to be content.
The Bridal Pair Robert W. Chambers *short story*	In this late Victorian ghost story, a man is united with his lost love, but not as he had expected.
The Question Pablo Neruda *poem*	The speaker expresses a passion and possessiveness that could be Heathcliff's.

Exploring the Connections

Making Meanings

Early Autumn

Connecting with the Novel

Though Hughes's story lacks the dramatic passion of Brontë's, how does it still capture the tragedy of a love unrealized?

1. Which character still seems in love with the other despite the passage of years? Explain what brought you to this conclusion.

2. Explain the **symbolic** significance of the title and the description of the setting.

3. List details in the story that convey how Bill and Mary's past relationship has affected their lives.

4. **The bus started. People came between them outside, people crossing the street, people they didn't know. Space and people. She lost sight of Bill.** How is Bill and Mary's parting **symbolic** of the things that divide people?

> **READING CHECK**
> Briefly summarize the events of the story.

If the Stars Should Fall
Sorrow Is the Only Faithful One

Connecting with the Novel

How does the first stanza of "Sorrow Is the Only Faithful One" relate to Heathcliff's life?

1. In the poem "If the Stars Should Fall," what effect does the repetition of the word *years* in the first stanza have on the tone of the poem?

2. What other words are repeated in "If the Stars Should Fall"? What is the effect of that repetition?

3. Samuel Allen uses stars and Owen Dodson uses mountains as poetic devices to convey sorrow. How are they used? Why are stars and mountains effective choices—more so than the moon or the sea would have been?

4. Explain how the speaker in "Sorrow Is the Only Faithful One" contrasts his sorrow and himself with stars.

> **READING CHECK**
> a. What does the speaker say he would do "if the stars should fall"?
> b. To what does the speaker compare sorrow in "Sorrow Is the Only Faithful One"?

78 | *Wuthering Heights*

Exploring the Connections (cont.)

Making Meanings

I see around me tombstones grey

Connecting with the Novel

Re-read Catherine's dream in Chapter IX and her statements to Linton after he enters her room in Chapter XII. What elements from Brontë's poems are also present in these passages?

1. What words make the strongest impact on you as you read through the poem? How do they affect you?
2. To whom does "they" in the poem refer?
3. Midway through the poem, the word "tenants" is used as personification. What is personified? Where are these "tenants" housed?
4. The speaker personifies Heaven and Earth. How does the Earth regard Heaven? What does that reveal about the **theme** of the poem?
5. Traditionally the earth is referred to as "mother"; how does Brontë extend that **metaphor** in the last part of the poem?

Reading Check
a. What pains are never healed?
b. Where does the speaker long to stay?

"Mr. Bell's" Wuthering Heights

Connecting with the Novel

The second review raises the issue of Heathcliff as hero. Is he a hero? Explain, including your definition of a hero.

1. In your opinion, does Heathcliff represent humanity in a "wild state," the state we all would experience if it were not for our "civilized training"? Explain.
2. Explain the **extended metaphor** of nature that the writer of the first review uses to launch his analysis of characterization in *Wuthering Heights*.
3. Both essays take issue with the author for dragging "into light all that he discovers, of coarse and loathsome [nature]." Does Brontë highlight the worst of human nature with her extreme characters? Explain.
4. Would these reviews have been more favorable or less had the critics been aware that Ellis Bell was actually Emily Brontë? Support your opinion, using the reviews and your knowledge of the social mores of the time as evidence.

Reading Check
a. What elements of the novel did the critics dislike?
b. Who do the critics assume to be the author of *Wuthering Heights*?

Exploring the Connections (cont.)

Making Meanings

"Reader, I Married Him"

Connecting with the Novel

In the light of this information about marriage customs of Victorian England, how do the characters' decisions about marriage become more understandable?

1. What are the advantages of marital customs like those of Victorian England? the disadvantages?
2. What is the difference between the marriage banns and the several types of marriage licenses?
3. Explain the link between marriage and social position.
4. List three changes in wedding etiquette between the Victorian Era and the present.

READING CHECK
a. What happened to a woman's property when she married?
b. What was pin money?
c. What recourse might you have if your betrothed broke the engagement?

Heston Grange

Connecting with the Novel

Locate passages from *Wuthering Heights* that reflect the feeling toward the Yorkshire area described in this selection.

1. From descriptions provided in this passage, what about Yorkshire would or would not appeal to you?
2. Helen Alderson and James Herriot eventually marry. What clues that theirs would turn into a long-lasting relationship are present in the passage?
3. What about the landscape of Yorkshire creates a symbolically significant setting for *Wuthering Heights*? Cite an example from this selection to illustrate your point.

READING CHECK
a. What is James Herriot's profession?
b. What brings him out to Heston Grange?

The Unquiet Grave

Connecting with the Novel

At what point in the novel are events described that mirror the events of "The Unquiet Grave"?

1. Do you find the devotion of the man admirable? Explain.
2. **Alliteration** is the repetition of the same or similar consonant sounds in words that are close together. Locate three examples of alliteration. Explain.
3. Why is the dialogue style effective for this romantic poem?
4. **Assonance** is the repetition of similar vowel sounds followed by different consonant sounds in words that are close together. Locate two examples of assonance.

READING CHECK
a. Who are the speakers in the poem?
b. What does the male speaker wish?
c. What advice does the female speaker give him?

Exploring the Connections (cont.)

Making Meanings

The Bridal Pair

Connecting with the Novel

What elements of this late Victorian ghost story echo the mood, theme, and style of *Wuthering Heights*?

1. Why is the idea of a lost love so appealing to audiences that writers of every generation return to it?
2. What is **ironic** about the theory upon which the young man is working and the events of the story?
3. What is significant about the locale and date of the young man's reunion with his love?
4. What details **foreshadow** the information revealed to the young man by the girl who seems to have followed him all over the world?
5. Compare and contrast the young man's appearance and circumstances of death with Heathcliff's at the novel's conclusion.

> **READING CHECK**
> a. What is the young man's occupation?
> b. Where did he originally meet the woman he loves?
> c. Where are the two when they are reunited?
> d. What does the man eventually discover about his love?
> e. In the conclusion, what becomes of the man?

The Question

Connecting with the Novel

If Heathcliff were the speaker of this poem, at what point in the novel would he have felt this way? Explain.

1. What, do you suppose, is the question referred to in the first two lines?
2. How does the speaker explain his absence?
3. What effect does a mundane word like *toenails* have upon the intensity of this poem?
4. Explain the **tone** (attitude of the speaker to the subject) and meaning of "I am not the passenger or the beggar, / I am your master. . . ."

> **READING CHECK**
> a. What is the desire of the speaker?
> b. How does the speaker feel he is treated by his beloved?

From "The Question" from *The Captain's Verses* by Pablo Neruda, translated by Donald D. Walsh. Copyright © 1972 by Pablo Neruda and Donald D. Walsh. Reprinted by permission of **New Directions Publishing Corporation.**

Study Guide | 81

Name _____ Date _____

TEST — PART I: OBJECTIVE QUESTIONS

A. Directions: Match each character with the correct description. Write the letter of the choice in the blank. *(2 points each)*

____ 1. Frances

____ 2. Nelly Dean

____ 3. Isabella

____ 4. Joseph

____ 5. Zillah

a. the wife of Heathcliff

b. the housekeeper at Wuthering Heights

c. the wife of Hindley

d. the housekeeper at Thrushcross Grange

e. a religious hypocrite

B. Directions: Circle the letter of the answer that best completes the statement. *(4 points each)*

6. *Wuthering Heights* claims to be the
   a. autobiography of Heathcliff
   b. diary of Mr. Lockwood
   c. memoirs of Catherine Earnshaw
   d. biography of Nelly Dean

7. Heathcliff acquires Wuthering Heights
   a. illegally
   b. according to Hindley's will
   c. by holding all the mortgages
   d. after Hindley dies without an heir

8. Of all the characters in the novel, Hareton most closely resembles
   a. his aunt, Catherine
   b. his father, Hindley
   c. his cousin, Cathy
   d. Nelly Dean

9. All of the following characters undergo significant changes in the course of the novel except for
   a. Catherine Linton
   b. Heathcliff
   c. Joseph
   d. Hareton

10. The continuing prescence of Catherine after her death is suggested through all the following means except
    a. Lockwood's dream at Wuthering Heights
    b. her own predictions
    c. Heathcliff's belief
    d. Lockwood's observations at her grave

C. Directions: In the space provided, mark each true statement *T* and each false statement *F*. *(2 points each)*

____ 11. Exactly how Heathcliff acquires his fortune during the years of his absence from Wuthering Heights remains a mystery.

____ 12. After she is kidnapped, Cathy escapes through the window of her mother's bedroom.

____ 13. After Heathcliff's death, Cathy intends to remain at Wuthering Heights.

____ 14. More than seventeen years after Catherine's death, Heathcliff opens her coffin and holds her one last time.

____ 15. Linton, Cathy, and Hareton all openly defy Heathcliff.

Name _____ Date _____

TEST — PART II: SHORT-ANSWER QUESTIONS

Directions: Answer each question using the line provided. *(3 points each)*

16. How does Heathcliff first come to reside at Wuthering Heights?

_____

_____

_____

17. How does Hindley react to his wife's death?

_____

_____

_____

18. Why does Nelly sit in when Edgar comes to Wuthering Heights to call on Catherine?

_____

_____

_____

19. Why does Heathcliff leave Wuthering Heights?

_____

_____

_____

20. Soon after Heathcliff moves back to Wuthering Heights, what changes take place in the upbringing of Hareton?

_____

_____

_____

Name _____ Date _____

 **PART II: SHORT-ANSWER QUESTIONS**

21. What do Heathcliff and Catherine quarrel about on the day she locks herself in her room, and what is the result?

22. Explain the circumstances around Isabella's escape from Wuthering Heights.

23. Why does Linton go to Wuthering Heights the day after he arrives at Thrushcross Grange?

24. What is the effect of Linton's fear of his father on his behavior toward Cathy?

25. What circumstances lead up to Hareton's throwing Cathy's books into the fire?

Name _____ Date _____

TEST — PART III: ESSAY QUESTIONS

Directions: Choose *two* of the following topics. Use your own paper to write two or three paragraphs about each topic you choose. *(30 points)*

a. Both Catherine Earnshaw and Cathy Linton marry for the wrong reasons. Explain the reasons behind each woman's decision. Who betrays herself more and why?

b. Describe the **point of view** that Brontë uses and what is unusual about it. Illustrate through reference to the text why this technique is effective.

c. What elements of the supernatural are found in the story? Which characters are affected by these elements? How are they affected?

d. Illustrate how the **internal** and **external conflicts** experienced by one of the characters are significant to the action of the story.

e. The story is set in three locations—Wuthering Heights, Thrushcross Grange, and the moors. Identify the **symbolic significance** of the locations, and explain how that symbolism contributes to the understanding of the characters.

f. Discuss how one of the **Connections** from the back of the novel (HRW Library edition) is related to a **theme, issue,** or **character** in *Wuthering Heights*.

Use this space to make notes.

Answer Key

Answer Key

Wuthering Heights

Chapters I–III

■ **Making Meanings**

> **READING CHECK**
> **a.** Lockwood is the tenant and Heathcliff is his landlord.
> **b.** Lockwood has rented Thrushcross Grange to escape the "stir of society" after he disappointed a young lady he adored by acting as though he didn't care about her.
> **c.** Lockwood learns that Mrs. Heathcliff is Heathcliff's widowed daughter-in-law and that Hareton is not Heathcliff's son, but an Earnshaw.
> **d.** Heathcliff refuses to provide a guide for Lockwood, and Lockwood fears that he would become lost and cold if he ventured forth alone.
> **e.** Lockwood learns that Catherine's brother Hindley mistreated Heathcliff as a child and refused to allow him to play with Catherine or eat with the family.

1. Responses will vary.
2. Lockwood is directly involved in the action. Responses will vary on reliability, but students should note that even though he does not immediately understand everything he witnesses, he is generally reliable as a narrator; he is a careful observer who is interested in understanding personalities and events.
3. Responses will vary but should include specifics from the text. Details of appearance include "black eyes," "dark-skinned gypsy," "rather slovenly," "erect and handsome." Personality details for Heathcliff include "exaggeratedly reserved," "surly," "morose," "very intelligent," and averse to "showy displays of feeling."
4. Outside, Wuthering Heights is buffeted by harsh winds that stunt the trees and erode the structure. Inside, the relationships between the inhabitants are harsh and stormy.
5. Instances of verbal irony include calling Mrs. Heathcliff an "amiable hostess"; describing Heathcliff and the two young people as "that pleasant family circle"; referring to their bickering as "civil behavior"; and labeling Joseph "my friend."
6. Responses will vary. Questions arising from suspense in the plot thus far may include the following: How did Heathcliff become the owner of Wuthering Heights when the name Hareton Earnshaw appears above the doorway? What is Heathcliff's connection to the present Hareton Earnshaw? Why does Heathcliff hate his daughter-in-law? Why is she not permitted to leave Wuthering Heights? Who is Catherine Earnshaw? Was the apparition at the window real?
7. Joseph is judgmental and uncharitable. He predicts Mrs. Heathcliff will go to the devil. He refuses to lend Lockwood a lantern that would help guide him in the storm.
8. First, Lockwood dreams that he is listening to a sermon in which 490 separate sins are discussed. He is beaten by Joseph and the rest of the congregation when he complains. In his second nightmare, he dreams he is awakened by a branch tapping against the window. When he breaks the glass to reach for the branch, his fingers close on those of a cold, small hand. A voice sobs, "Let me in," and identifies itself as Catherine Linton, home after twenty years. He scrapes the arm of the apparition against the broken window glass, causing it to bleed. It seems doubtful he believed they were more than dreams: He refers to Heathcliff's emotional display as "a piece of superstition."
9. Responses will vary. Love, hate, and revenge are often reasons that deceased characters in stories haunt the living.

Answer Key (cont.)   *Wuthering Heights*

10. Heathcliff is both generous and callous. Heathcliff offers Lockwood wine and a place by the fire; however, he laughs at his guest when Lockwood is attacked by the dogs. Heathcliff is rude to members of his own household in front of a guest. He is inhospitable—he will not provide a guide for Lockwood to Thrushcross Grange or appropriate lodging for the night as an alternative. Students will probably note that hospitality is still respected and expected today.

■ Reading Strategies Worksheet
Characterization
Chart: Responses will vary.

- Lockwood appears the perfect example of urban sophistication, but his antics with Heathcliff's dogs reveal him as somewhat silly.
- Hareton appears lacking in manners, social skills, and education, but he displays a natural pride when introducing himself to Lockwood.
- Mrs. Heathcliff seems very scornful and uncaring, but she expresses concern to Heathcliff that no one will help Lockwood get home.

Follow-up: Responses will vary, but should reflect attention to Lockwood's traits as well as to the personality and values of the character who is describing those traits.

Chapters IV–IX

■ Making Meanings

> **READING CHECK**
> a. Nelly Dean takes up the narration in Chapter IV.
> b. On a trip to Liverpool, Mr. Earnshaw found Heathcliff abandoned and starving and brought the young child home to Wuthering Heights.
> c. Mr. Earnshaw favors him above his own children; Mrs. Earnshaw ignores him; Hindley resents and finally dominates him; Catherine becomes his best friend.
> d. Hearing them laughing, a bulldog takes after the pair and bites Catherine on the ankle, bringing her down. While she is brought inside the Grange to be cared for, Heathcliff, because of his dark looks, is told to go home.
> e. Catherine is in fine clothes, beautifully groomed, and displaying good manners.
> f. A drunken Hindley dangles baby Hareton over the banister. When the baby falls, Heathcliff instinctively catches him.
> g. Catherine explains to Nelly that in marrying Edgar she would gain social standing, whereas marriage to Heathcliff would degrade her. She says she could use Edgar's money to protect Heathcliff.

1. Responses will vary.
2. Frances grows easily out of breath and has a troublesome cough. She says she is afraid of dying. She becomes almost hysterical after seeing people dressed in black, a sign of mourning.
3. Wuthering Heights is a comfortable farmhouse situated on a bleak hilltop surrounded by desolate moors. Thrushcross Grange sits in a sheltered valley surrounded by a park. Wuthering Heights has durable oak furniture, a stone floor, and beamed ceilings, while Thrushcross Grange has crimson carpets, upholstered chairs and sofas, gilt ceilings, and crystal chandeliers. While the inhabitants of Wuthering Heights exhibit violent passions, those of Thrushcross Grange are reasonably genteel.

Answer Key (cont.) *Wuthering Heights*

4. Hindley humiliates Heathcliff on Christmas Eve by forcing him to appear dirty and ungroomed in front of Catherine. The next day, after Heathcliff has improved his appearance, Hindley ridicules his "elegant locks" and shuts him in the garret to prevent him from joining the Christmas festivities. Heathcliff resolves to make Hindley pay for his cruelty. Rather than feel pain he plots revenge.

5. Edgar Linton is Heathcliff's foil. As Lockwood discerns from Edgar's portrait, Linton was soft-featured, pensive, amiable, light-haired, and "almost too graceful." He is afraid of Hindley and shies away from violent confrontations. He is rich and exhibits an air of social superiority. Heathcliff is the exact opposite: hard-featured, moody, unsociable, dark-haired, graceless, fearless, poor, and socially inferior.

6. Hindley resents Heathcliff because Mr. Earnshaw obviously prefers him over his own son. Hindley brutally gains revenge after his father's death. Heathcliff, though once a street-urchin, has a fierce pride and strong will, but his resistance to Hindley grows into hatred and desire for revenge, especially when he hears Catherine say she cannot marry him because Hindley has degraded him.

7. Responses will vary. The storm might symbolize Catherine's violent passions. The tree might represent Heathcliff split apart by Catherine's decision to marry Edgar, or the division in her own soul over the two men, each seeming an embodiment of one side of her personality.

8. Responses will vary. Some students may find this emotion overly dramatic and too possessive to be romantic; other may find it in keeping with emotions expressed in love songs and popular movies. Students should distinguish between romantic and realistic expectations of love.

9. Responses will vary but should reflect familiarity with the text.

10. Responses will vary. Students should consider how Nelly feels about a character when she makes judgments about that character's behavior. Students may think it is fair because it reflects real life.

■ Reading Strategies Worksheet

Cause and Effect

Box 1 Heathcliff's absence from dinner and disappearance.

Box 2 Catherine develops a fever after sitting up all night in wet clothes.

Box 3 The Lintons catch her fever.

Final Outcome: The Lintons die within days of each other.

Follow-up: Responses will vary. Students should explore the idea that neither Heathcliff nor Catherine wants to hurt the other and that they focus on each other to the exclusion of everyone else.

Answer Key (cont.)

Wuthering Heights

Chapters X–XVII

■ **Making Meanings**

> **READING CHECK**
> a. Catherine greets him with joyous enthusiasm. Edgar greets him with haughty resentment, calls him a "plough-boy," and suggests he be entertained in the kitchen.
> b. Catherine tells Isabella that Heathcliff will marry her only for her money, then crush her.
> c. Possible responses include the following: Heathcliff lends Hindley large sums of money to encourage his further decline, teaches Hareton to curse his father, deprives Hareton of education, and assumes financial control over Wuthering Heights.
> d. Heathcliff has contempt for Isabella.
> e. Heathcliff prays Catherine's spirit will wake in torment and that her ghost will haunt him and drive him mad.
> f. After Heathcliff throws a knife that strikes Isabella behind the ear, she flees to Thrushcross Grange for her possessions, then settles in London.
> g. Because of Hindley's gambling debts to him, Heathcliff, rather than Hareton, inherits the Earnshaw property.

1. Responses will vary.
2. He has changed. He is militarily erect. He has a full beard and foreign-sounding voice.
3. Isabella had complained that she is frequently dismissed from the company of Catherine and Heathcliff and reprimanded Catherine for her jealousy; in retaliation, Catherine wants to make a point of using Isabella's argument against her and therefore, without dismissing her, Catherine tells Heathcliff of Isabella's romantic interest in him. Catherine also wishes to dispel any thoughts Isabella might have that Heathcliff is a gentle and caring man, or that he would be interested in her.
4. Heathcliff realizes he could possess Thrushcross Grange through marriage to Isabella should Edgar and Catherine fail to produce a male heir.
5. When Edgar threatens to have Heathcliff forcefully removed from the Grange, Heathcliff derides him. When Catherine joins in the abuse and prevents her husband from summoning assistance, Edgar loses his nerve. However, when Heathcliff proceeds to bully him, Edgar strikes back and runs to summon help, but Heathcliff escapes before he returns. The incident causes Catherine to resent both Heathcliff and her husband for breaking her heart and brings about the onset of her brain fever. Catherine blames Edgar more than Heathcliff because she believes she could have persuaded Heathcliff to leave Isabella alone if Edgar had not interfered.
6. He despises Edgar for being Catherine's husband and says Isabella must suffer for him until Heathcliff can take his revenge on Edgar directly.
7. Catherine and Heathcliff embrace passionately, but Heathcliff is pained to see Catherine's deteriorated condition. She accuses him of killing her and wishes they were both dead. He feels tortured and accuses her of being possessed by a devil. She says she will not find peace in death, forgives him, and then asks his forgiveness. He hesitates but ultimately succumbs to her pleas. As they embrace again, Catherine faints. Later that night she gives birth to a daughter and dies without fully regaining consciousness.
8. Responses will vary. Hindley was self-destructive and weak, easily swayed by his wife and the vengeful Joseph. His treatment of Heathcliff created an enemy; his excessive drinking and poor self-control increased his violent tendencies. He used poor judgment in his choice of drinking and gambling partners, including Heathcliff, whose gambling wins enabled him to move into Wuthering Heights and begin his quest for revenge.

Study Guide | **91**

Answer Key (cont.) — *Wuthering Heights*

9. Students should realize that Heathcliff will not triumph over Edgar as easily as he did over Hindley. Hindley and Edgar are basically opposites, as Nelly notes, despite their "similar circumstances." Hindley alienates his child after the death of Frances, while Edgar finds solace in his child. Hindley is erratic, mean-spirited, and irresponsible; Edgar is constant, good-natured, and dutiful.

10. Responses will vary. Students should note Brontë's clues regarding both a supernatural and a psychological haunting by Catherine. Her "spirit" will continue to affect Heathcliff and the direction of his vengeance. Catherine's traits will reemerge in her blood relations: Cathy and Hareton.

■ Reading Strategies Worksheet

Summarizing and Drawing Conclusions

The First Quarrel

Conflicts: Catherine says Heathcliff should not pursue Isabella because he is not interested in her. Catherine says he hurts her when he hurts Edgar.

Heathcliff says Catherine treated him horribly by marrying Edgar, and he wants to hurt Edgar. He needs no one's approval to pursue someone.

Outcome: Attempting to stop the quarrel, Nelly alerts Edgar, who enters and joins the fray.

The Second Quarrel

Conflicts: Edgar criticizes Catherine's tolerance of Heathcliff's behavior, asks Nelly to get servants to throw Heathcliff out of the house, strikes Heathcliff, and leaves to get reinforcements.

Catherine calls Edgar a sneak and a coward.

Heathcliff ignores Edgar's orders to leave, taunts Edgar by knocking over his chair and threatening him, leaves before Edgar returns.

Outcome: Catherine, followed by Nelly, goes upstairs.

The Third Quarrel

Conflicts: Edgar demands that Catherine choose between him and Heathcliff.

Catherine demands to be left alone and begins banging her head against the sofa until she falls into a stupor.

Outcome: Edgar is terrified until Nelly tells him it is an act and to ignore her.

Follow-up: Paragraphs should include some of the following points:

Catherine's goals:

- Heathcliff directs his anger and desire for revenge at Edgar instead of at Catherine. Understanding this, Catherine believes she can control Heathcliff and persuade him to leave Isabella alone.
- Catherine intends to manipulate Edgar to get her way so that she can continue to see Heathcliff.

Edgar's goals:

- Edgar wants peace and a return to normalcy and will force Catherine to choose between him and Heathcliff to regain them.

Chapters XVIII–XXV

■ Making Meanings

> **READING CHECK**
> a. Initially they get along well, though Cathy refuses to believe Hareton is her cousin and treats him as if he were a servant. Hareton curses her but soon tries to appease her by offering her a puppy.
> b. Cathy is Hareton's cousin through the Earnshaws: her mother, Catherine, and Hareton's father, Hindley, were sister and brother. Cathy is Linton's cousin through the Lintons: Her father, Edgar, and his mother, Isabella, were brother and sister.
> c. He stays overnight. Then, he travels to Wuthering Heights with Joseph.
> d. Cathy and Linton meet on her sixteenth birthday.
> e. They will marry, paving the way, should Linton die, for Heathcliff to gain control of Thrushcross Grange.
> f. She likes him and Edgar forbids her to visit.
> g. When Nelly falls sick and her father is still ill, Cathy has the opportunity to escape for visits to the Heights.

Answer Key (cont.) *Wuthering Heights*

1. Responses will vary.
2. Nelly greatly admires Cathy but disliked her mother. Students may note that Nelly's attitude toward Cathy tends to make readers sympathize with her even when her behavior resembles her mother's.
3. Hareton curses Cathy after she mistakes him for a servant; to her surprise and disbelief, she learns he is her cousin.
4. Some details that may cause readers to be anxious for Linton include his weakness in the face of Heathcliff's strength; his unawareness of the existence of his father; his resemblance to Edgar, whom Heathcliff hates; his reaction to the exterior of Wuthering Heights.
5. Hareton shows her around the stables because Linton does not respond to his father's request and does not wish to exert himself.
6. Cathy and Linton have little, if anything, in common. She is generous; he is selfish. He prefers idleness; she loves rambling on the moors. Except for his lack of polish, Hareton's nature more closely resembles hers. Like her mother before her, who chose the superficialities of a conventional marriage over deeply passionate love, Cathy risks tragedy by preferring Linton. The difference is that Cathy does not yet love Hareton, whereas her mother did love Heathcliff. Cathy may be denying her nature, but she is not betraying her heart.
7. Heathcliff tells Cathy that Linton is dying of love for her. Cathy cries and tells Nelly she wants to judge for herself what the situation is.
8. Linton, who is feverish, says the kiss will take his breath away. This reaction is indicative of his nature, which does not give generously or receive generously.
9. Accept reasonable responses.
10. Responses will vary. Most students will probably assert that literacy is just as important today, and that in most countries citizens are presumed to be literate. Students may respond additionally that a person's means of expression, written or spoken, does influence the impression he or she makes.

■ Reading Strategies Worksheet

Responses will vary but should include the following points.

Comparing

Cathy has her father's fair complexion, light hair, and sympathetic, kind, and forgiving nature. Cathy has her mother's dark eyes, her indomitable spirit, and sense of loyalty.

Follow-up: Answers will vary depending on the scene chosen.

Chapters XXVI–XXXIV

■ Making Meanings

> **READING CHECK**
> a. Linton is lying on the ground; he appears weak and in declining health.
> b. Heathcliff locks Cathy and Nelly in the house, giving Cathy, in her mind, little choice but to marry Linton in order to return to her father.
> c. Linton leaves the bedroom door unlocked.
> d. His unkempt appearance reminds Heathcliff of himself when he was young; he looks like his Aunt Catherine.
> e. Cathy at first avoided Hareton's company. When she was forced by the cold to join the others by the fire, she continually scorned his attempts at friendship.
> f. He says he has done no injustices and repents nothing.
> g. Lockwood discovers a changed house: flowers are blooming, Heathcliff has died, and Hareton and Cathy are planning to marry.

1. Responses will vary.
2. Responses will vary but students should be aware of the following: Catherine is revealed as a can-

Study Guide | 93

Answer Key (cont.) — *Wuthering Heights*

did, strong-willed, sensible young woman, who is selfless—agreeing to marry Linton if it will enable her to return to her father. Heathcliff's ragings have robbed Linton of any sense of honor or dignity. He is a shivering, sobbing invalid virtually immobilized by fear. Heathcliff is cruel; the height of the cruelty is seen in this chapter when he berates his son and throttles Catherine after making her and Nelly prisoners. Nelly is an observer, as she has been all along, neither preventing nor consciously assisting Heathcliff's plans.

3. By removing Edgar's emotional support, Heathcliff hopes to hasten his death. Even after Cathy marries Linton, Heathcliff cannot be certain of success as long as Edgar lives, because Edgar is both moral and wise.

4. Edgar orders Nelly to send for the lawyer so that he can protect Cathy's personal inheritance by putting her money in trust for her and her children. Though Nelly sends for the lawyer, he does not come until Edgar is dead because he is working for Heathcliff.

5. Heathcliff has twice attempted to open Catherine's coffin. Heathcliff tells Nelly that he persuaded the sexton to open Catherine's coffin so that he could look at her face. Seeing her again fills him with peace. On the night of her funeral, he tried to dig up her corpse and embrace her, but he stopped when he heard sighing and became convinced that Catherine was with him. In the intervening years, he has been constantly tortured by feeling her presence but not being able to see her.

6. When Heathcliff sees Cathy and Hareton reading together, he admits to Nelly that he has lost the will for revenge because all the enjoyment is gone. Heathcliff's "strange change" involves his seeing Catherine everywhere, especially in Hareton. Though he feels healthy, he anticipates the fulfillment of the "one universal idea" that has devoured his existence—being with Catherine again.

7. Responses may include that Cathy escapes through the window, plants a flower garden without Heathcliff's permission, and continues to show affection for Hareton despite Heathcliff's warnings. Students might say that Heathcliff is unable to subdue Cathy because of the character traits she inherited from her parents: loyalty, capacity for forgiveness, and perseverance. Linton had none of these traits; Hareton does not have his aunt's fiery will.

8. Heathcliff yearns to achieve two things: revenge, and peace at Catherine's side. By relenting in his desire to achieve the first goal, he is able to liberate himself from earthly attachments—the hell in which he lived—and direct his energies toward attaining the second.

9. The blooming flowers and neat appearance signify that the occupants are now calm and even-tempered and that Wuthering Heights is a pleasant place to live.

10. Responses will vary. Students should consider Hareton's treatment under his father's care as one reason for his affection for Heathcliff.

11. Responses will vary but should be based upon the text. Students may feel Lockwood could have been excluded, allowing Nelly to act as primary narrator; yet to whom would she be telling her tale? Zillah makes a brief appearance as narrator; though the information could have been conveyed in a different way, this allows Nelly to be a gatherer of information like Lockwood and the reader.

■ Reading Strategies Worksheet

Problem and Solution

Responses may vary.

Circumstance/Obstacle: Heathcliff wants Linton and Cathy to marry, but it is difficult for him or Linton to

Answer Key (cont.) — *Wuthering Heights*

persuade her to marry if she is not allowed on the Heights property.

Means of overcoming obstacle: He lures Nelly and Cathy there with the promise of a ride, placing Linton, too ill to ride, just off Thrushcross Grange property. With each visit to Linton, Nelly and Cathy are lured closer to Wuthering Heights.

Circumstance/Obstacle: Heathcliff wants Cathy to come to the Heights so he can force her to marry Linton.

Means of overcoming obstacle: Heathcliff kidnaps Nelly and Cathy, tells the group from Thrushcross Grange that arrives later that day that Cathy and Nelly are not there, and the next morning, forces Cathy to marry Linton but will not release her from the grounds.

Circumstance/Obstacle: Heathcliff is concerned that the dying Edgar might rewrite his will.

Means of overcoming obstacle: He bribes the lawyer, Mr. Green, so that he does not change Edgar's will.

Follow-up: Responses will vary. Students will obviously note the lengths to which Heathcliff will go to acquire that which he desires; they should also note his psychological astuteness and persistence in developing a scheme: he is aware of Cathy's desire to be with her father and obey him and uses that to his advantage; and he is willing to wait, and let his son die, to achieve his aim.

Literary Elements Worksheets

■ Symbolism

Responses will vary. Following are possible responses:

Symbol: "On that bleak hill-top the earth was hard with a black frost, and the air made me shiver through every limb." Chapter II, page 7 (HRW LIBRARY)

Conflict/Emotion: Mr. Lockwood's feeling of being unwelcome and his perception that Wuthering Heights is a cold and forbidding place are symbolized in this passage.

Symbol: "The rainy night had ushered in a misty morning, half frost, half drizzle, and temporary brooks crossed our path. . . ." Chapter XXIII, page 243 (HRW LIBRARY)

Conflict/Emotion: This is symbolic of the cool reception that Cathy and Nelly will receive from Linton at the Heights, as well as the temporary argument that will cross the path of Linton and Cathy.

Symbol: "It was a close, sultry day: devoid of sunshine, but with a sky too dappled and hazy to threaten rain. . ." Chapter XXVI, page 268 (HRW LIBRARY)

Conflict/Emotion: The weather the afternoon that Linton and Cathy meet on the heath symbolizes the reunion—uncomfortable and unhappy, but not so much so that it cancels the afternoon visit.

Follow-up: Responses will vary.

■ Multiple Narrators

Responses will vary but should include some of these points.

Chapters I–III: Lockwood: Like Lockwood, the reader is new to the situation and characters and experiences them with the same curiosity and frustration as Lockwood.

Chapters IV–IX: Lockwood, Nelly, Heathcliff: The reader encounters the characters and the setting at the same pace as Lockwood, and this creates suspense; Nelly Dean's narration is more subjective, because she not only retells events but colors them with her own interpretation. Heathcliff's account of the visit to the Lintons is one of the few times in the novel his point of view is presented.

Chapters X–XVII: Nelly, Isabella Linton: Nelly's narration is sympathetic to Edgar, rather critical of Catherine, and generous to Heathcliff; Isabella pro-

Answer Key (cont.) — Wuthering Heights

vides the reader with insight into events that paint a negative portrait of Heathcliff.

Chapters XVIII–XXV: Nelly, Cathy Linton: Because Nelly can move freely between both the Grange and the Heights, she provides the information that lets the reader see the contrast between these two households; Cathy's perspective is refreshingly naive and wholesome.

Chapters XXVI–XXXIV: Nelly, Lockwood, Zillah, Linton, Heathcliff: As the story draws to a close, these narrative voices tie up all the loose ends and allow the reader to realize the depth of despair and depravity to which Heathcliff has sunk and his release from those burdens in death; and the effect of Heathcliff's cruelty on Cathy and Linton—Linton's comments and Zillah's narration in particular provide insight into this. Lockwood's narration presents the change in Wuthering Heights and its inhabitants from hostile and dark to lively and light.

Follow-up: Responses will vary. Students should be aware of the limitations of these narrators, their admitted prejudices and shortsightedness; however, they should also address the characters' positions in the novel that allow them to be privy to information that is vital to the development of the story.

■ Indirect Characterization

Responses will vary but should include some of these points.

Heathcliff's words: He refers to Edgar as a "puny being." He calls his wife "a mere slut!" He refers to his own son as "tin" instead of gold and as a "whey-faced whining wretch." He calls his daughter-in-law worthless and a "damnable witch!"

Heathcliff's actions: Heathcliff strikes Edgar and Cathy, pushes Nelly, kicks Hindley, and throws a knife at Isabella; in contrast, he rushes to see Catherine when he can, waits outside while Catherine is dying, replaces Edgar's hair in her locket with his, asks the sexton to remove the sides of his and Catherine's coffins so that their dust can mingle.

Nelly Dean: Nelly is kind to Heathcliff; she tries to include him when he is younger with Earnshaw family activities; she helps him improve his appearance; and she allows him access to Catherine before she dies. Despite all she knows of him, her words about him are more understanding than harsh.

Catherine Earnshaw: There is continual contradiction between Catherine's words and actions. Catherine considers Heathcliff to be her soul mate, yet she marries Edgar Linton.

Isabella Linton: At the beginning of her infatuation with Heathcliff she declares to Catherine, "I love him more than ever you loved Edgar . . ." She runs off with him despite the fact that he tried to hang her dog. At the end of the relationship she tells Nelly Heathcliff is a "Monster! . . . [h]e's not a human being. . . ." She runs away to London after she finally escapes from the Heights.

Catherine Linton: Cathy is at first pleased to meet her uncle; ultimately this pleasure turns to disgust: "You are miserable, are you not? Lonely, like the devil, and envious like him? Nobody loves you—nobody will cry for you when you die."

Heathcliff's character: Heathcliff is cruel and selfish except to Catherine. He shows little regard for anyone else and is held in slight regard by others. Nelly allows the reader to see some sympathetic sides to his character; his interaction with Cathy and Isabella reveals his tyrannical nature.

Follow-up:
- Successful responses will include the above information in paragraph form.
- Responses will vary. Successful answers will take into consideration words and actions by characters, and when appropriate the narrator of the account. For instance, though Isabella and Catherine parted on poor terms, Isabella indicates by her mourning that she loved her sister-in-law; Catherine appears to have loved Isabella as well, and the advice she

Answer Key (cont.)

Wuthering Heights

gave her was sound, but the manner in which she dealt with her reveals her selfish nature.

■ Foreshadowing

1. foreshadows that Heathcliff's return will cause problems at Thrushcross Grange and for Edgar and Catherine
2. foreshadows that the passion between Heathcliff and Catherine will be renewed
3. foreshadows that Cathy will leave the grounds and betray Nelly's trust
4. foreshadows that Cathy was disobeying Nelly and traveling secretly to Wuthering Heights

Follow-up: Responses will vary. Successful student responses will include the quotation, circumstances around the passage that prompted it, and what event it foreshadows.

■ Gothic Literary Elements

Responses will vary. Following are possible responses.

Scene: Chapter XII, page 130–131 (HRW Edition)

Gothic elements: Catherine's hallucinations when she has locked herself in a room at Thrushcross Grange and "sees" Wuthering Heights and Heathcliff and talks about their being joined in death.

Scene: Chapter XXIX, page 298–299 (HRW Edition)

Gothic elements: Heathcliff tells Nelly he tried to open Catherine's coffin to embrace her again and felt her spirit pass him, a spirit that he says has haunted him for eighteen years.

Scene: Chapter XXXIV, page 348 (HRW Edition)

Gothic elements: Ghosts of Heathcliff and Catherine walking on the moors are seen by the shepherd boy.

Follow-up: Responses will vary.

Vocabulary Worksheets

If you wish to score these worksheets, assign the point values given in parentheses.

■ Vocabulary Worksheet 1

Chapters I–IX

A. *(4 points each)*

1. d. ill-tempered
2. a. untidy
3. c. silent
4. b. sound judgment
5. d. firm
6. b. blameworthy
7. a. composure
8. b. stubbornly
9. d. strong dislike
10. c. gloomy
11. a. objecting earnestly
12. d. intruder
13. b. arrogantly
14. b. bewildered fear
15. a. ill will

B. *(4 points each)*

16. b. robust
17. e. complaining
18. c. gloomy
19. h. charming
20. g. generous
21. d. dull
22. a. bully
23. j. revealed
24. f. imperceptible to the touch
25. i. irritably

■ Vocabulary Worksheet 2

Chapters X–XXXIV

A. *(4 points each)*

1. a. indirect remarks
2. d. just beginning
3. c. anxiety
4. a. agree
5. c. lack of doubt
6. d. sluggishness
7. b. weak and thin
8. a. idleness
9. c. warning
10. b. extended

B. *(4 points each)*

11. a. pessimistic
12. b. provoked
13. d. touchable
14. b. apparent
15. c. flexible

Answer Key (cont.)

Wuthering Heights

C. *(4 points each)*

16. d. started again
17. g. took possession of
18. b. arrogant
19. f. habitual
20. j. ideal
21. a. summary
22. e. crept sideways
23. h. to make flesh
24. c. cleverly
25. i. greed

Exploring the Connections

■ Early Autumn

> **READING CHECK**
> Bill and Mary, who were formerly in love, meet on a street in New York City. They briefly catch up with each other and she suggests having dinner so their respective spouses and families can meet. Her bus arrives, Mary gets on, and then realizes they did not exchange addresses and that she also neglected to tell him that her youngest boy is named Bill.

1. Students will probably draw the conclusion that Mary is still in love with Bill; Bill's first thought on encountering his former love is simply that she looks old.
2. Autumn is traditionally a time of change. Nature shuts down gradually, and vegetation begins to die. The title is indicative of the season, but also perhaps of the premature death of their relationship when "something not very important had come between them," or of the abrupt conclusion to their reunion. Leaves fall without a wind, symbolic of the change that occurs, with little excitement or emotion, at this meeting. Mary realizes she is no longer young and the past is out of reach. The New York street lights are "misty," "blurred," and twinkle, perhaps like Mary's recollection of their love.
3. Responses will vary. Students should note that Bill turned bitter toward women as a result of Mary's decision to marry another man. Mary has named one of her children after Bill.
4. Responses will vary. Students may note that time, distance, and other factors can cause people to lose sight of one another.

Connecting with the Novel

Mary's choice of another man over him turned Bill bitter toward women; certainly Heathcliff's life is marked with this bitterness as well. There is something very desperate in the woman's reaction to seeing her former love. Even though she has married someone else, she has named her son after this man. Just as Heathcliff married someone besides Catherine and had a child with this wife, yet never stopped loving Catherine, this character has never stopped loving Bill.

■ If the Stars Should Fall
Sorrow Is the Only Faithful One

> **READING CHECK**
> a. The speaker would "grant them privilege to fall." In other words, he would not object.
> b. He compares sorrow to a season, the laughter of an enemy, and that which clings like erosion scars to a cliff.

1. Repetition of *years* establishes a tone of monotony and slow passage of time.
2. Words that are also repeated are *less, care, same, cold, all,* and *down.* These words are stark and monotonous, which seems to be the speaker's experience with life.
3. Stars and mountains are symbolic of enormous and fixed elements of nature. They do not appear to change, and neither does the experience of sorrow expressed in these poems. Since the moon and the sea change, they would not have been effective symbols.
4. The stars the speaker describes are reflections and therefore without life. The speaker makes a distinction between these stars and his sorrow. His sorrow is real and enduring unlike a reflection in water.

Answer Key (cont.) — *Wuthering Heights*

Connecting with the Novel

Heathcliff clings to sorrow to the extent of pushing all others away so that sorrow can be his only companion for the long years after Catherine's death. It is his "original skin": he was found an orphan, starving on the streets of Liverpool; he was not welcomed by Mrs. Earnshaw or Hindley at Wuthering Heights; and he is rejected by Catherine for Edgar.

■ I see around me tombstones grey

> **READING CHECK**
> a. "Time, and Death, and Mortal pain/Give wounds that will not heal again."
> b. The speaker would prefer to stay on earth even if in a grave.

1. Responses will vary.
2. "They" refers to those who live in heaven, "sweet land of light," angels perhaps.
3. The "tenants" are "[t]orments and madness, tears and sin!" that dwell in the human condition, haunting "each mortal cell."
4. Earth turns away from Heaven. One of the themes of the poem is preference for earth, even with its pain, over an afterlife elsewhere.
5. Brontë extends the metaphor of earth as mother by describing people as children who don't want to leave her even to go to a dazzling place; who want to see her face as long as they can; and who want their "lasting rest" to be on her.

Connecting with the Novel

Catherine, like the speaker of the poem, would rather be on earth, even in a tomb, than in heaven. In Chapter IX, she describes a dream of being in heaven, yet miserable. "[H]eaven did not seem to be my home; and I broke my heart with weeping to come back to earth. . . ."

■ "Mr. Bell's" *Wuthering Heights*

> **READING CHECK**
> a. The critics felt the novel showed negative aspects of human character that did not need to be revealed; they describe the characters as violent and grotesque.
> b. The critics assume Mr. Ellis Bell is the author.

1. Responses will vary.
2. The critic likens the behavior of characters in *Wuthering Heights* to phenomena that, while natural, are also grotesque and frightening. To the reviewer, the characters and circumstances of *Wuthering Heights* are "scenes of savage wildness . . . which though they inspire no pleasurable sensation, we are yet well satisfied to have seen."
3. Responses will vary but should be based upon the text.
4. Students who feel reviews would have been less favorable may think that because women were considered "the gentler sex," Brontë's exploration of such dark subjects and personalities would be considered inappropriate. The second reviewer already feels modern novels are full of "affectations and effeminate frippery," which he might have applied to *Wuthering Heights* had he been aware the author was female. Students who feel the reviews would have been more favorable may feel that critics would have been sympathetic to a first-time female novelist.

Connecting with the Novel

Responses will vary based upon the students' definition of a hero. Some students may argue that Heathcliff's tyrannical rule over everyone in his life makes it difficult to think of him as heroic. He does no deeds of valor, which many associate with heroism. However, other students may find him a very sympathetic lead character, for whom they find them

Answer Key (cont.) — *Wuthering Heights*

selves rooting. By the novel's conclusion many will see the character as redeemed.

■ "Reader, I Married Him"

> **READING CHECK**
> a. A husband assumed possession of his wife's property upon marriage.
> b. Pin money was an allowance to a wife.
> c. A broken engagement was like a broken business contract; the injured party could bring a lawsuit.

1. Responses will vary. Students may see advantages in the level of stability and security the institution of marriage offered during that era, and they may appreciate the straightforward, businesslike quality of these unions. Possible disadvantages that students might raise include the lack of equality for women, pressure to marry within or above one's social class, and the husband's burden of complete responsibility for the financial stability of the family.

2. Banns, notice of the intent to marry, were published, allowing someone to oppose the union in its first three months. If the union was unopposed during that period of announcement, the marriage was valid thereafter. There were three types of licenses: one to marry in a parish to which the couple belonged; a special license allowing one to marry in any parish at any time; and a civil license, obtained from a government clerk, for a couple not associated with the Church of England.

3. Not only did marriage allow one to improve one's social position by marrying "up," it also afforded one the chance to maintain position in society and often maintain property with the receipt of a handsome dowry. The posting of banns was a lower-class custom; a special license signaled privilege.

4. Three changes from the etiquette of the Victorian Era include the timing of the ceremonies, which no longer need to be conducted before 3:00 P.M; the reception, for which a breakfast is no longer expected; and the honeymoon, on which a bride no longer brings a chaperone.

Connecting with the Novel

Responses will vary. (1) Catherine Earnshaw's reasons for marrying Edgar seem more realistic in light of this information. (2) Students may not regard Edgar Linton quite so harshly. He does attempt to protect Cathy from Heathcliff's designs on her property. Though not happy about his daughter's plans to marry Linton, Edgar is being realistic in recognizing it is the only way for her to keep the Grange.

■ Heston Grange

> **READING CHECK**
> a. James Herriot is a veterinarian.
> b. He has come to Heston Grange to tend a lame calf.

1. Responses will vary.

2. Students may note that the following points foreshadow a lasting relationship for the couple: Both Alderson and Herriot appreciate the Yorkshire Dales; Helen is at ease with the calf and running a farm, which would be appealing to a veterinarian; Herriot feels immediately companionable with Helen, which he says is unusual for him.

3. The nature of the romantic relationship between Heathcliff and Catherine is similar to the description of Yorkshire. For example, "a lot of people find it . . . wild. It almost seems to frighten them."

Connecting with the Novel

Responses will vary. In Chapter IX Catherine explains to Nelly that she would rather be on the heath than in heaven. Her daughter, in Chapter XVIII, is captivated by the moors and the cliffs and the birds and flowers that flourish there.

Answer Key (cont.) — *Wuthering Heights*

■ The Unquiet Grave

> **READING CHECK**
> a. The speakers are a living man and his dead love.
> b. The male speaker wishes that he could kiss his dead love.
> c. She tells him that to kiss her would hasten his death, and that he should be content with life.

1. Responses will vary.
2. Three examples of alliteration are "For I **c**rave one **k**iss of your **c**lay-**c**old lips," "**g**arden **g**reen," and "**f**inest **f**lower."
3. Responses will vary. The dialogue creates a sense of the pair as still united despite death.
4. Two examples of assonance are "sp**ea**k . . . w**ee**ping . . . sl**ee**p," "sl**ee**p . . . s**ee**k."

Connecting with the Novel

Heathcliff recounts to Nelly his attempt to open Catherine's grave after her death and hold her one last time. This is interrupted by an overwhelming feeling of her presence, which comforts Heathcliff. Upon Edgar's death, he again considers opening her coffin, but satisfies himself by having the sexton loosen the coffin side so eventually the dust of their bodies can mingle.

■ The Bridal Pair

> **READING CHECK**
> a. The young man is a doctor.
> b. He originally met her in elementary school.
> c. The story is set in the woods, on a hill above a cemetery.
> d. He discovers that his love is a ghost.
> e. The man dies in the end.

1. Responses will vary.
2. The young man is studying insanity; he is looking for an organic source in the body for mental illness. When Rosamund explains that she is dead, he recognizes symptoms of mental disturbance with which he is familiar; he becomes excited about discovering a cure for her.
3. The reunion is exactly three years to the day that the man saw and fell in love with the mysterious young woman. He discovers that the first vision of her on the hill above the cemetery was the day of her death.
4. The young man mentions that he has been in love only once in his life and that was when he was ten; her name was Rosamund, and he believed she had passed away recently. He boards a train named *Rosamund* for his vacation. He stays in cabin thirteen during both this hunting trip and his previous trip, three years earlier.
5. His physical condition deteriorates to the point that other characters comment that he looks "tired to death." As with Heathcliff, there seems to be no medical reason for such a young man to die. After the young man's death, as with Heathcliff's, "at first they thought he was asleep."

Connecting with the Novel

The moods are similar, especially the aura of the supernatural created at the outset of *Wuthering Heights* and established in "The Bridal Pair" by the girl's omnipresence in the young man's life. The theme of the novel is repeated in the story: a couple have loved each other since childhood and have been separated by death. The man, like Heathcliff, is desperate to be united with his love; death is no obstacle for their love. Both emphasize destiny. Both are written in the complex prose style of the Victorian Era and reflect an interest in the Gothic.

■ The Question

> **READING CHECK**
> a. The speaker desires to be let fully back into the life of his beloved.
> b. She keeps him at a distance with a shadow between them that he does not want.

Answer Key (cont.)

Wuthering Heights

1. Responses will vary but will probably be confined to a variation of "Do you love me?"

2. He went away in "thorny uncertainty," but the roads he traveled had all been waiting for him, though he had not known it.

3. Responses will vary. The inclusion of *toenails* on the list of things the speaker loves seems to add to the totality of his adoration.

4. Responses will vary, but the tone, despite the use of the word *master*, is not so much domineering as passionate and intense toward the beloved. The speaker refuses to be dismissed; he is to be the central consideration of her life.

Connecting with the Novel

This seems an attitude that Heathcliff would have adopted upon his return after three years away. He discovers that Catherine has married while he has been working to make himself worthy of her; he has come back "from thorny uncertainty." He virtually takes up residence at the Grange and in her heart again, and ultimately she cannot cope with that.

Test

■ Part I

1. c
2. d
3. a
4. e
5. b
6. b
7. c
8. a
9. c
10. d
11. T
12. T
13. F
14. F
15. F

■ Part II

Responses will vary but should include the following points.

16. Mr. Earnshaw, the father of Catherine and Hindley, returns from Liverpool with Heathcliff and raises him as a son.

17. He begins to drink heavily and associate with a bad crowd. He also becomes abusive and violent at home.

18. Hindley has asked her to act as chaperone.

19. He hears Catherine say she will marry Edgar and that it would degrade her to marry Heathcliff.

20. Heathcliff dismisses Hareton's tutor and plans to use the boy as an instrument of revenge against Hindley.

21. Catherine wants Heathcliff to stop pursuing Isabella because she knows he does not really love her. After a fight breaks out between Heathcliff and Edgar, Catherine locks herself in her room and grows delirious from lack of food.

22. Hindley wrestles Heathcliff to the ground after Heathcliff throws a knife at her, cutting her neck beneath the ear. Isabella then runs away.

23. Heathcliff sends Joseph for Linton. When Edgar says he will be brought in the morning, Joseph warns that Heathcliff will come for him if they delay further.

24. Linton's fear of his father causes him to trick Cathy and Nelly into returning to Wuthering Heights, where the women are held until the young couple are wed; to delay helping Cathy escape to see her dying father; to remain silent during Heathcliff's verbal attacks on Cathy, because they directed Heathcliff's attention from him.

25. She continues to scorn his attempts to read and be friends.

■ Part III

Responses will vary but should include the following points.

a. Catherine agrees to marry Edgar because he is rich. She believes she loves him, and that she could use his money to help Heathcliff escape him. Catherine knows she loves someone else, so she betrays her own heart by agreeing to marry Edgar.

Cathy agrees to marry Linton at Heathcliff's insistence. She wants to return to her father before he

Answer Key (cont.)

Wuthering Heights

dies. She believes she loves Linton, and so students may feel that her self-betrayal is not as severe.

b. Brontë uses the first-person point of view with multiple narrators. Usually books told in first person have one narrator. Students may note the following advantages: the point of view contributes to the suspense of the novel as the story is slowly revealed and the reader receives different insights from different characters about an event, just as one would in real life.

c. The novel contains references to ghosts and hauntings. Catherine and Heathcliff are most affected. Both see visions of the other. Catherine seems to survive as a ghost, waiting for Heathcliff to join her in death.

Notes

Notes

Notes

Notes

Notes

Notes